Acacias

Acacias: The Genus Acacia (sensu lato) is an evidence-based treatment of this supergenus, through the eyes of a clinical pharmacognosist and integrative medicine specialist. The book begins with antiviral activity, revealing within the five genera of *Acacia s.l.* pharmacological properties and pharmacologically active compounds. Profiles of prominent species within these genera, including photographs, accompany the narrative of current research and traditional usage into their antibacterial, antifungal, anticancer, antidiabetic, metabolic syndrome ameliorative, and psychotherapeutic potential.

Features

- Comprehensive treatment of the entire *Acacia sensu lato* genus
- Aids ethnopharmacological prospectors in new sources of novel botanically based medicines for modern metabolic and psychiatric diseases
- Illuminates the presence of psychedelic simple substituted tryptamines in trees and their medical and psychotherapeutic potential

Continuing in the tradition of the previous volumes of the esteemed Medicinal and Aromatic Plants – Industrial Profiles series, *Acacias: The Genus* Acacia (sensu lato) provides a unique and comprehensive coverage of one of the most interesting and diverse genera of trees, firmly entrenched in the Levant, Africa, Australia, the Far East, and the New World. The influence of these genera on pharmacy and industry (especially through gum arabic, wildcrafted in Africa from *Senegalia senegal*), human consciousness, the advent and development of religions, planetary ecology, botanical therapeutics, and the emergence of psychedelic medicine reflect both the history of our species and the transformative promise of tomorrow.

Medicinal and Aromatic Plants—Industrial Profiles

Individual volumes in this series provide both industry and academia with in-depth coverage of one major genus of industrial importance.

Series Edited by
Ephraim Shmaya Philip Lansky

Volume 26
Citrus, *edited by Giovanni Dugo and Angelo Di Giacomo*

Volume 27
Geranium and Pelargonium, *edited by Maria Lis-Balchin*

Volume 28
Magnolia, *edited by Satyajit D. Sarker and Yuji Maruyama*

Volume 29
Lavender, *edited by Maria Lis-Balchin*

Volume 30
Cardamom, *edited by P.N. Ravindran and K.J. Madhusoodanan*

Volume 31
Hypericum, *edited by Edzard Ernst*

Volume 32
Taxus, *edited by H. Itokawa and K.H. Lee*

Volume 33
Capsicum, *edited by Amit Krish De*

Volume 34
Flax, *edited by Alister Muir and Niel Westcott*

Volume 35
Urtica, *edited by Gulsel Kavalali*

Volume 36
Cinnamon and Cassia, *edited by P.N. Ravindran, K. Nirmal Babu, and M. Shylaja*

Volume 37
Kava, *edited by Yadhu N. Singh*

Volume 38
Aloes, *edited by Tom Reynolds*

Volume 39
Echinacea, *edited by Sandra Carol Miller. Assistant Editor: He-ci Yu*

Volume 40
Illicium, Pimpinella and Foeniculum, *edited by Manuel Miró Jodral*

Volume 41
Ginger, *edited by P.N. Ravindran and K. Nirmal Babu*

Volume 42
Chamomile: Industrial Profiles, *edited by Rolf Franke and Heinz Schilcher*

Volume 43
Pomegranates: Ancient Roots to Modern Medicine, *edited by Navindra P. Seeram, Risa N. Schulman, and David Heber*

Volume 44
Acacias: The Genus *Acacia (sensu lato)*, *edited by Ephraim Shmaya Philip Lansky, Helena Maaria Paavilainen, and Shifra Lansky*

Acacias

The Genus *Acacia* (*sensu lato*)

Ephraim Shmaya Philip Lansky
Helena Maaria Paavilainen
Shifra Lansky

CRC Press
Taylor & Francis Group
Boca Raton London New York

CRC Press is an imprint of the
Taylor & Francis Group, an **informa** business

Designed cover image: Zipora Lansky

First edition published 2023
by CRC Press
6000 Broken Sound Parkway NW, Suite 300, Boca Raton, FL 33487–2742

and by CRC Press
4 Park Square, Milton Park, Abingdon, Oxon, OX14 4RN

CRC Press is an imprint of Taylor & Francis Group, LLC

Library of Congress Cataloging-in-Publication Data
Names: Lansky, Ephraim P., editor. | Paavilainen, Helena M., editor. | Lansky, Shifra, 1991- editor.
Title: Acacias : The Genus *Acacia* (*sensu lato*) / Edited by Ephraim Philip Lansky, Helena Maaria Paavilainen, and Shifra Lansky.
Description: First edition. | Boca Raton, FL : CRC Press, 2023. | Includes bibliographical references and index. | Summary: "Acacias: The Genus *Acacia* (*sensu lato*) is an evidence-based treatment of this super genus, through the eyes of a clinical pharmacognosist and integrative medicine specialist. The book begins with antiviral activity, revealing within the five genera of *Acacia s.l.*, pharmacological properties and pharmacologically active compounds. Profiles of prominent species within these genera, including photographs, accompany the narrative of current research and traditional usage into antibacterial, antifungal, anticancer, antidiabetic, metabolic syndrome ameliorative, and psychotherapeutic potential"— Provided by publisher.
Identifiers: LCCN 2022045956 (print) | LCCN 2022045957 (ebook) | ISBN 9781138339804 (hardback) | ISBN 9780429440946 (ebook)
Subjects: LCSH: Acacia. | Medicinal plants. | Materia medica, Vegetable.
Classification: LCC QK495.M545 A23 2023 (print) | LCC QK495.M545 (ebook) | DDC 583/.633—dc23/eng/20220927
LC record available at https://lccn.loc.gov/2022045956
LC ebook record available at https://lccn.loc.gov/2022045957

ISBN: 978-1-138-33980-4 (hbk)
ISBN: 978-1-032-46966-9 (pbk)
ISBN: 978-0-429-44094-6 (ebk)

DOI: 10.1201/9780429440946

Typeset in Times
by Apex CoVantage, LLC

For Professor Benny Shanon,
who brought the nighttime to light

E.S.P.L.

For Robynne, Sygal, Lena
– friends indeed!

H.M.P.

Contents

List of Structures

Preface

This Preface was written before the outbreak of the COVID-19 pandemic, which has since eclipsed the threat of AIDS as the viral scourge uppermost in the public consciousness and imagination. In a strange way, the pandemic also gave us time and space to finish this project.

This book focuses on the pharmaceutical and modern health potential of trees of the genera formerly known as *Acacia*. The trees of this group were in recent years found to be polyphyletic and not monophyletic as formerly believed, and after some debate, it was decided by a congress of professional botanists to split the former grouping into several genera, the original *Acacia* covering most but not all of the trees in Australia and some from Asia, the Pacific Islands, and the Mascarene Islands (*vide* Bruce Maslin, 2015, *Gardens' Bulletin Singapore* 67: 231–250). Although some of the other genera are now also found in Australia, most are in Africa, the Middle East (*Senegalia* and *Vachellia*) and the New World (*Acaciella* and *Mariosousa*), and as a common name, "acacia" is still in wide use throughout the world. Therefore, the genus name *Acacia* may be adopted for the entire group with the appendage *s.l.*, or *sensu lato*, i.e., "in the wider sense." This book uses the term "acacias" as the common name for these trees worldwide, noting the controversy and reasons that all these trees, once thought to be of one genus, are now not considered so.

Overall, the emphasis herein encompasses three major and overlapping perspectives, specifically: (1) ethnomedical uses, (2) chemistry, and (3) pharmacology of principal components and pharmaceutical potential. Since ancient times, the resin exuded by one (or more) of the African species, i.e., *Senegalia senegal*, has been employed by pharmacists and physician-pharmacists as a binding agent and emulsifier widely known as "gum arabic." The gum, rich in glycoproteins and polysaccharides, may potentiate the immunomodulating potential of other agents and has also been shown to be immunomodulating in its own right. Other trees in this group, such as the species presently known as *Senegalia catechu*, suppress replication of the HIV-1 virus (associated with Human Immunodeficiency Syndrome) through viral protease and associated viral proteins. Thus, as a group, this diverse and polyphyletic grouping of trees exhibits important potential for the treatment of diverse viral diseases, especially those manifesting with immunological suppression.

Saponins are a large group of amphiphilic compounds found in many plants. Because of their ability to enter both aqueous and lipid compartments, they have important medical uses both as potential primary drugs and as immunological adjuvants. Their amphiphilic nature may also facilitate blood-brain barrier penetration. For example, distinctive triterpenoid *Acacia* (*s.l.*) saponins, the acaciasides, are remarkable both for their antifilarial (killing parasitic worms) actions and their ability to improve the effectiveness of tetracycline and other modern antimicrobials. These unique *Acacia* (*s.l.*) saponins also possess spermicidal activity, though they lack mutagenic potency. Other characteristic saponins found in selected species of *Acacia* (*s.l.*) include the prospogenols and concinnosides (Abul Gafur et al. 1997) with putative anodyne effects as well as their better-known use, for example from the species *A. concinna*, in cleansing the hair when incorporated into shampoos, primarily in India from antiquity. Saponins are of great importance, not only for making soaps and detergents but also as immunological adjuvants used even today in modern vaccines (e.g., from the soapbark tree, *Quillaja saponaria*).

One species of *Acacia* (*s.l.*), i.e., *Vachellia tortilis*, including its single-stem variety, *V. tortilis* ssp. *raddiana*, still often referred to as *Acacia raddiana*, is widespread in the deserts of the Middle East, including the Israeli Negev and Saudi Arabia. The leaves, which have been reported to contain dimethyltryptamine (DMT) (Strassman 2000) as well as likely saponins, were recently claimed to exert experimental antidepressant and anxiolytic effects in mice (Alharbi and Azmat 2016). Because of the many serious side effects attached to current pharmaceutical drugs designed for treating depression and anxiety, the discovery of these activities in a common Middle Eastern acacia is highly encouraging as a potential source of pharmaceutical (including complex pharmaceutical) leads. The treatment of depression and anxiety has become a huge socio-economic problem in the modern world, presenting great challenges particularly when depression and anxiety are comorbid. Therefore, the discovery of potential treatment solutions for anxiety and depression in this plant with minimal toxicity and known activity against diabetes, cough, fungal and other infectious diseases, malaria, diphtheria, leishmaniasis, hypercholesterolemia, and inflammation suggests new vistas for potential medical applications and heuristic avenues for future research (Yadav et al. 2013). Discovering how different types of compounds, like monoamine alkaloids and saponins, might interact is a challenge that will help us learn more about the complex phytopharmaceutic activity of the plant.

ACACIAS AND HIV

Another important aspect to be specifically considered in this study is the discovery that compounds, especially proteins, found in acacia products may be of value for treating patients with acquired immunodeficiency syndrome (AIDS). The functionality relates both to the activity of the compounds against the replication of the human immunodeficiency virus and to their effect in the treatment of diseases to which the patients, with their compromised immunity, are particularly susceptible. Here is a brief outline of this subject:

1. Activity of *Vachellia nilotica* pod extracts against HIV proteases
2. Use of acacia gums in the preparation of non-soluble vaginal rings for novel delivery
3. Anti-HIV activity by *V. tortilis* and *Senegalia confusa* stem bark extracts
4. HIV reverse transcriptase is stopped by proteins from *S. confusa* seeds called acafusin and acaconin
5. Potential botanical-pharmaceutical interactions between acacia extracts and conventional anti-retroviral drugs
6. Coinfection in AIDS patients by *Leishmania* trypanosomes and other opportunistic parasites was halted by *V. nilotica* and other African species

RESEARCH NOTE: BENNY SHANON AND *ACACIA* (*S.L.*)

Acacias: The Genus Acacia (sensu lato) may be most properly considered the sequel of our third book in the CRC series Traditional Herbal Medicines for Modern Times, namely, *Harmal: The Genus* Peganum. The reason for this is largely due to the efforts of Professor Benny Shanon, Professor Emeritus at the Hebrew University of Jerusalem. Professor Shanon's field is cognitive psychology, and for several decades he has devoted his own considerable cognitive faculties to exploring the phenomenology of individual human reactions to the South American medicinal and visionary brew known as Ayahuasca. How this came about and how this relates to *Acacias* will now be considered.

As a cognitive psychologist, Professor Shanon became an expert in the categorization of human experience. Along the way, he became exposed to Ayahuasca in South America and later became highly experienced in its use. He maintained careful and detailed notes of his own sessions with the brew. Further along, he applied the same arduous scientific rigor to documenting not only his own Ayahuasca experiences but also those of others. The others included Western entheo-tourists who flew to South America to undergo the often difficult psychophysiological experience, native South American shamans, and native South American individuals who were not shamans. A tremendous range of highly complex experiences was reported, including conversing with beings from other dimensions, traveling to magnificent jeweled cities with highly elaborate architecture, and discoursing with animals, plants, and spirits.

In short, ayahuasca refers to a blend of herbs, one of which is the liana *Banisteriopsis caapi*, and the second is any of a number of plants that are rich in dimethyltryptamine (DMT), most notably *Psychotria viridis*.

Shanon calls the visionary vine *B. caapi* the "engine of ayahuasca" for its ability to maintain the psychoactive ground for the Ayahuasca effect to occur. This natural bioengineering is responsible for the *in situ* synthesis of the so-called harmala alkaloids, especially harmine, harmaline, tetrahydroharmine, and harmol. When the chemistry of *B. caapi* was first studied, one isolated compound looked most likely to be responsible for the psychic phenomena attributed to the plant's use, and it was dubbed "telepathine" by those early researchers. Later, it was finally realized that the mysterious telepathine was none other than harmine, a compound already known for its presence in *harmal*, i.e., *Peganum harmala*, especially in that plant's seeds.

Acacia s.l. enters into this discussion because DMT has been isolated from parts such as stem pericarp (bark), if not seed pods and leaves, from a number of its different species. Shanon, who had established himself as a major academic authority on ayahuasca, also came forward a few years back with a "speculative hypothesis" that an "ayahuasca analogue" might have been employed by ancient Hebrews as an intentional or accidental inebriant through the combining of some sort of brew containing a source of harmala alkaloids, namely, *P. harmala*, with a DMT-containing plant, possibly an acacia indigenous to the Middle East, such as *Vachellia tortilis* (Shanon 2008). Like other writers, such as R. Gordon Wasson (2021), who speculated that psychoactive plants or fungi may have given birth, through the induction of mystical experiences facilitated by psychoactive compounds which the plants or mushrooms contained, to the very idea of deity, Shanon takes the same great leap in considering the roots of Judaism and the great religions, i.e., Christianity and Islam, which it spawned, as possibly having been watered with similar psychoactive potions.

Such metaphysical and religious discussions are perhaps best kept beyond the scope of the present text, but the impact of such pharmacologically induced "entheogenic" experiences, insofar as they apply to medicine and pharmacognosy, is germane. As a result, the possibility that acacias contain dimethyltryptamine and that they have been used as entheogens in the past and present is intriguing.

What, indeed, is the relationship of inebriation with ayahuasca or its analogues to medicine? As was discussed in our previous CRC book, *Harmal*, the harmala alkaloids, such as the lead compound, harmine (which is actually ubiquitous as a signaling molecule throughout the entire plant kingdom), exert potent tonifying and balancing effects not only in the nervous systems of animals but on their immune systems as well. Of over three thousand compounds screened in a large Harvard University/National Institutes of Health investigation, only harmine exhibited the properties of lowering global body inflammation and causing the regeneration of insulin-secreting pancreatic islet beta cells. As such, the possible use of this compound or plants that contain it as a treatment for type 1 (juvenile onset) diabetes mellitus, generally thought incurable, was duly noted.

Also of note was the presence of serotonin receptors on the cell surfaces of both neurons and lymphocytes, the latter being the putative "cognitive" component of the immune system. It was also thought that the nervous and immune systems, which are both affected by serotonin, might work together as a "syncytium," and that substances like harmine, which are known to affect serotonin receptors, might have psychosomatic effects not only from the brain to the body, but also from the body to the brain through "lymphoneuric" immuno-neural circuits.

Serotonin, as well as its derivative, melatonin, are biological signaling molecules that respond to perturbation, and they are found throughout the animal kingdom and, in at least some circumstances, in plants. Commonly thought of as neurotransmitters, a larger context involving both animals and plants is needed for describing the effect of these compounds on the organism, whether within the animal or plant kingdoms.

Serotonin is built upon the indole structure, which is identical to the chief growth hormone in plants, i.e., indole. The indole alkaloid tryptamine is the direct precursor to serotonin and may also transform into various endogenous hallucinogens, including possibly the harmala structure 10 methoxyharmalan (Lansky 1975) as well as DMT. In fact, DMT is found both in animal and plant tissues in multiple examples. Why are these powerful psychoactive compounds found in plants as well as in animals, including ourselves, who *produce*, or can produce, DMT and harmala congeners in response to certain *situations*?

So there you have it. What kinds of *situations* would require a plant to have the ability to synthesize within itself DMT? The conventional assumption was always that the DMT provided the plant with a chemical defense against herbivory—that its potential predators would become too inebriated to remember the plant's location or even what they were doing in the first place. However, plants also have a kind of "immune system" involved in their own protection against injury and infection, and injury may take on a very broad meaning when considering "psychic" trauma as well as physical trauma. If plants also have the intelligence to find the best points to anchor vines and other such maneuvers, they are likely traumatized from personal injury in both the physical and "psychic" dimensions. And what about human beings? What benefit to the individual or the species does the ability to produce DMT confer? Why can humans do this? For what purpose? And when?

If DMT, which apparently does occur in at least some species of *Acacia s.l.*, at least some of the time, may, as Strassman and others have suggested, occur in humans at the moments before death and provide a kind of preparation for the transition, what type of function does it provide for a plant?

Many of the Ayahuasca experiences described by Shanon and others involve communication between the "soul" or essence of a plant and that of the human. In short, the experiences describe a kind of telepathy between animal (i.e., human) and plant life. In animals and humans, it is related to the interaction of its endogenous neural indolic chemistry with the indolic hallucinogens in plants, such as those found in the Peruvian and Brazilian rainforests and in *Acacia s.l.* trees throughout the world. And for the trees? What do they "get out" of it? Is it a way that plants transport themselves to other worlds? Or communicate with each other over vast expanses of space? Of time?

ACACIAS AND THE SHAMANIC QUEST

In a recent communication to the journal *Clinical Toxicology (Philadelphia)*, Liu and colleagues (Liu et al. 2019) described two cases of "N,N-dimethyltryptamine (DMT) poisoning" in two male roommates in their early twenties who had consumed together an "herbal stew" consisting of seeds from the *Peganum harmala* plant (Lansky et al. 2017) boiled together in water with bark from an unidentified tree of "acacia." This report, the first in the medical literature to describe a clinical case of interaction between the harmala alkaloids found in *P. harmala* and substituted tryptamine(s) contained in an *Acacia* (*s.l.*) plant part, is notable for more than one reason.

First, the authors point out that the combination of the two plants is apparently designed to mimic the effect of the combination of the harmala alkaloid-containing bark of the South American liana *Banisteriopsis caapi* and the DMT-containing leaves of the South American shrub *Psychotria viridis*, the two plants being the usual main components of the legendary entheogenic brew known as Ayahuasca, which, as they indicate, has recently become very popular with young adults worldwide. The symptoms of poisoning they describe included alteration of consciousness, elevation of liver enzyme activity, and rhabdomyolysis, a breakdown of muscle tissue sometimes associated with drug abuse, for example, heroin or cocaine. They dismiss the alkaloids in *P. harmala* as monoamine oxidase (MAO) inhibitors, which allow the "toxic" DMT to act and cause "sympathomimetic" damage.

Pharmacologically speaking, the authors are at best only partially correct in their assessment. While *P. harmala* is widely known as a poisonous plant (in high doses), acacias do not generally have such a reputation. Each year, there are deaths recorded from the ingestion of *P. harmala* seeds, which in small doses are valuable and extremely popular (for example, in Pakistan and Iran) medicine for many health conditions. Virtually all the deaths attributed to *P. harmala* each year are due to overdosing, which most commonly occurs in women attempting to abort a fetus they are carrying with the known abortifacient. On the contrary, fatalities attributed to acacias of any species or of any of their parts are, if they occur at all, at least extremely rare. So, it is possible that the changes in consciousness that are part of the toxicity syndrome are caused by DMT, but it is more likely that the potentially life-threatening changes in the body are caused by the *P. harmala* and its beta-carboline alkaloids, such as harmine, harmaline, tetrahydroharmine, harmol, and harmane. However, the deeper question hidden herein is: for what reason did those two young men decide to prepare and ingest such a brew, and similarly, what is the reason that many other young people worldwide are now experimenting with Ayahuasca or with "Ayahuasca analogues" (Ott 1994) which may sometimes include extracted parts, usually leaves or stem barks, of acacias as a source of DMT? To better understand this phenomenon, it may be worthwhile to temporarily leave the desert habitats common to acacias and return to the South American rainforests. In such settings, the quest for expanding the mind finds a cultural context for its expression. It might be argued that the inclination to alter one's consciousness is at the root of a deep, collective, and evolutionarily conserved impulse with an apparent competitive advantage for the "traveler" (Saniotis 2010; Sullivan and Hagen 2002).

REFERENCES

Abul Gafur, M., T. Obata, F. Kiuchi, and Y. Tsuda. 1997. *Acacia concinna* saponins. I. Structures of prosapogenols, concinnosides A-F, isolated from the alkaline hydrolysate of the highly polar saponin fraction. *Chem Pharm Bull (Tokyo)* 45: 620–5.

Alharbi, W.D., and A. Azmat. 2016. Pharmacological evidence of neuro-pharmacological activity of *Acacia tortilis* leaves in mice. *Metab Brain Dis* 31: 881–5.

Lansky, E.S., S. Lansky, and H.M. Paavilainen. 2017. *Harmal: The Genus Peganum*. Boca Raton, FL: CRC Press.

Lansky, P. 1975. *Consciousness and the Ergotropic and Trophotropic Systems of Arousal*, Baccalaureate Thesis, New College, Sarasota, FL.

Liu, C.H., W.L. Chu, S.C. Liao, C.C. Yang, and C.C. Lin. 2019. Syrian rue seeds interacted with acacia tree bark in an herbal stew resulted in N,N-dimethyltryptamine poisoning. *Clin Toxicol (Phila)* 57: 867–9.

Ott, J. 1994. *Ayahuasca Analogues: Pangæan Entheogens*. Kennewick, WA: Natural Products Company.

Saniotis, A. 2010. Evolutionary and anthropological approaches towards understanding human need for psychotropic and mood altering substances. *J Psychoactive Drugs* 42(4): 477–84.

Shanon, B. 2008. Biblical entheogens: A speculative hypothesis. *Time Mind* 1: 51–74.

Strassman, R. 2000. *DMT: The Spirit Molecule: A Doctor's Revolutionary Research into the Biology of Near-Death and Mystical Experiences*. Rochester, VT: Park Street Press.

Sullivan, R.J., and E.H. Hagen. 2002. Psychotropic substance-seeking: Evolutionary pathology or adaptation? *Addiction* 97(4): 389–400.

Wasson, R.G. 2021. *Soma: Divine Mushroom of Immortality; Ethno Mycological Studies*. [s.l.]: Lulu.com. See also at www.bookdepository.com.

Yadav, P., R. Kant, and P. Kothiyal. 2013. A review on *Acacia tortilis*. *Int J Pharm Phytopharmacol Res* 3: 93–6.

Acknowledgments

E.S.P.L.: I wish to thank my co-authors, Helena Paavilainen and Shifra Lansky, for their excellent and patient work on this book. I am indebted to Zipora Lansky (ziporalanskyart.com) for her evocative watercolor of the lone *Acacia s.l.* tree in the desert moonlight. Dr Ori Fragman-Sapir provided invaluable assistance for the field identification of *Acacia s.l.* trees in Israel. As always, I am grateful to Professor Eviatar Nevo, Founder of the Institute of Evolution at Haifa University, for his many kindnesses, model, and context.

H.M.P.: My heartfelt gratitude goes to Ephraim Shmaya Philip Lansky and Shifra Lansky. The whole idea of the book came from Ephraim, and his creativity shines through on every page. But without Shifra's painstaking and detail-oriented work, the project could never have been completed. You are a wonderful team. I also want to thank Dr. Elaine Soloway for her encouragement and kindness in taking me on a private field trip in the Negev to teach me about Israeli acacias. I am grateful to Randy Brehm for her support and never-ending patience during the multiple delays in the delivery of the manuscript. And, last but not least, my great support team: Prof. Shmuel Kottek, whose advice and constant encouragement have carried me through my whole career; Robynne Bruckenstein, Sygal Amitai, and Lena Tiemeyer, friends who have kept joy and hope alive even during the corona months; and my mother and sister, Maija and Kaarina Paavilainen, who are always there for me, even at moments of failure. Thank you for making my work meaningful.

Authors

Ephraim Shmaya Philip Lansky holds an MD from the University of Pennsylvania Perelman School of Medicine and a PhD in pharmacognosy from Leiden University. He is licensed in Israel as both a physician and a hypnotherapist, and currently, during the pandemic, he practices classical homeopathy and Hypnoidal Suggestive Therapy remotely with patients worldwide over Zoom. Dr. Lansky is a well-known expert in the field of Medical Punicology, the use of pomegranate fruit, *Punica granatum*, for cosmeceutical and medical treatments, and is the co-author of three books by CRC Press on the genera *Ficus*, *Capparis*, and *Peganum*. He has authored or co-authored over 35 scientific papers and is a Research Fellow at the University of Haifa in the Institute of Evolution, an interdisciplinary biological think tank and research center. He is also Series Editor for CRC's two series on medicinal plants, *Medicinal and Aromatic Plants: Industrial Profiles* and *Traditional Herbal Medicines for Modern Times*, both series inaugurated and established by Dr. Roland Hardman. Dr. Lansky welcomes book proposals for either of these series from potential authors.

Helena Maaria Paavilainen is a researcher at the Hadassah Medical School, Hebrew University of Jerusalem, Israel. Her main research interests are ethnomedicine, historical ethnopharmacology, and the history of pharmacology, especially the Hebrew, Arabic, and Latin traditions. She wrote her PhD thesis (published as *Medieval Pharmacotherapy: Continuity and Change; Case Studies from Ibn Sina and Some of His Late Medieval Commentators*, Leiden: Brill, 2009) on the development of medical drug therapy in medieval times and on the potential validity of medieval herbal treatments. She also co-authored with Ephraim Lansky the monograph *Figs: The Genus Ficus* (Boca Raton, FL: CRC Press, 2010), and with Ephraim and Shifra Lansky the sequels *Caper: The Genus Capparis* (2014) and *Harmal: The Genus Peganum* (2017). On the historical side, she has written a series of articles on the therapeutic choices of medieval physicians, for example, "Therapeutic Approaches in the Writings of Isaac Israeli," in *Isaac Israeli: The Philosopher Physician*, eds. K. Collins, S. Kottek, and H. Paavilainen (Jerusalem, 2015, pp. 139–171).

Shifra Lansky holds a PhD in chemistry from the Hebrew University of Jerusalem in Israel and is currently working as a postdoctoral fellow at Weill Cornell Medicine in New York. Her research focuses on studying the structure and dynamics of various membrane proteins. Dr. Lansky co-wrote the books *Caper: The Genus Capparis* and *Harmal: The Genus Peganum* with Ephraim Shmaya Philip Lansky and Helena Maaria Paavilainen (CRC Press). She has also written 19 peer-reviewed articles for international scientific journals.

1 Antiviral Effects of Acacias

Note: Structures of chemical compounds mentioned in the text can be found with the help of the List of Structures in the preliminary part of the book.

At the time of this writing, viruses are on everyone's minds, so it is apropos to begin this exploration of *Acacia s.l.* with its potential to modify, attenuate, and subdue the global virulence and attendant isolation in which we find ourselves. Acacias are the oldest trees on the planet and have seen it all. What part of their complicated knowledge of pharmacognosy could be used to help them fight the modern plague?

There is an old pun once delivered by the poet of blessed memory, Alan Ginsberg. It goes like this:

Question: Is life worth living?
Answer: It depends on the liver.

Indeed, from a biochemical point of view, all mammalian life may be said to descend from the Hepatic Function. According to Chinese medicine, liver is associated with the element Wood, the archetype of all green plants and the only living phase within that five-element system. The other elements, or phases—fire, earth, metal, and water—are active and imbued with the life force of the universe but not alive in the way of wood. The health of the liver indeed shapes one's love of life.

One of the most common and important diseases of the liver is its pathological inflammation, i.e., hepatitis. Most of the different types of hepatitis, or hepatitides, are caused by viruses, especially the hepatitis viruses A, B, and C. Extracts of sixty or more Saudi Arabian medicinal plants were screened for their potential virucidal action against the hepatitis B virus (HBV). Ethanolic extracts from two *Acacia s.l.* species, specifically from the leaves of *Senegalia mellifera* (Arbab et al. 2015) and *Vachellia oerfota* (Arbab et al. 2017), were among the best antivirals against HBV. People thought that the virucidal effects and the ability to stop the hepatitis B virus from making antigen in a lab dish were caused by mixtures of alkaloids, tannins, flavonoids, and saponins.

Still on the hepatic theme, the Hepatitis *C* virus (HCV) is also associated with significant morbidity and mortality in Africa and throughout the world, with a severe paucity of options for its therapy. In an *in vitro* assay, an N-butanol methanolic extract of *Acacia confusa* seeds resulted in inhibition of the RNA replication of HCV (Lee et al. 2011).

In another study involving the hepatitis C virus, the viruses were inoculated into human liver cells. Using a methanolic extract of undisclosed plant parts of *Vachellia nilotica*, a 50% reduction of viral growth in the model was achieved (Rehman et al. 2011).

An additional study showed that extracts of *V. nilotica* were effective against the virus causing the potentially fatal disease among goats and sheep, ovine rinderpest (which does not infect humans). The antiviral activity of extracts of the leaves was greater than that of extracts of the pods. In this case, extracts of the tree's bark had no effect against the virus (Raheel et al. 2013). However, extracts of the same tree were able to stop the spread of the goatpox virus in sheep at the highest safe dose (Bhanuprakash et al. 2008).

Dengue fever is a tropical disease in humans caused by an RNA virus, the Dengue fever virus (DENV). Though individuals usually recover, the disease may be fatal, particularly in infants and if the disease has progressed to the stage of "hemorrhagic fever." Like most viral diseases, treatments beyond fluid maintenance and rest are scarce, and new therapies are in fierce demand. The disease

DOI: 10.1201/9780429440946-1

(a) (b)

FIGURE 1.1 *Senegalia mellifera* is commonly known as blackthorn, since the lumber becomes jet black after oiling. The species name *mellifera* means sweetness and refers to both the pleasing aroma of the blossoms and their attraction for bees. (a) *S. mellifera* subsp. *detinens*, buds and white blossoms. (8/25/2018, Ghanzi, Botswana: by Robert Taylor, CC BY 4.0, via iNaturalist, www.inaturalist.org/photos/53186396.) (b) *S. mellifera* branch with leaves and thorns. (4/30/2021, Ngamiland West, North-West, Botswana: by Robert Taylor, CC BY 4.0, via iNaturalist, www.inaturalist.org/photos/126691846.)

(a) (b) (c)

FIGURE 1.2 *Vachellia oerfota*, the *orfot*, is an African and Middle Eastern species well known in Iran, where its seed has been analyzed and recognized as a valuable source for a high-protein drink (Zarei et al. 2015). Its value for the ecosystem as a "rare shrub with restricted distribution" has been both appreciated in the Kingdom of Saudi Arabia (Alatar et al. 2012) and vilified as an "encroaching species" due to "anthropogenic" environmental disruption in Ethiopia (Fetene et al. 2016). (a) *V. oerfota* var. *oerfota* flower and thorn. (8/4/2020, near As Sahwah town, Al Qunfudhah, Saudi Arabia: by Ali Mohammed Alzahrani, with permission.) (b) *V. oerfota* var. *oerfota* branch with seed pods and leaves. (3/25/2016, Ajyād, Makkah Province, Saudi Arabia: Ali Mohammed Alzahrani, with permission, via Flickr, www.flickr.com/photos/90607335@N07/49741280423/.) (c) *V. oerfota* var. *oerfota* branches with flowers, buds, and leaves. (8/4/2020, near As Sahwah town, Al Qunfudhah, Saudi Arabia: by Ali Mohammed Alzahrani, with permission.)

is endemic to at least 22 African nations and 120 nations worldwide. In one study, extracts of 33 different Thai plants were tested for possible antiviral activity against DENV. Panya et al. (2019) found that an extract from *Senegalia catechu* had the most antiviral activity among them.

Another virus against which *Acacia s.l.* species have been studied is the human immunodeficiency virus (HIV), the causative agent of acquired immunodeficiency syndrome (AIDS), still epidemic throughout Africa. HIV is an RNA retrovirus, meaning that it inserts its genetic material into the host cell, which it uses to produce for itself its very own DNA, the reverse of the usual process, hence the use of the term "retrovirus."

(a)

(b)

FIGURE 1.3 *Acacia confusa*. Sometimes known as Formosan acacia, this Southeast Asian native is known both for its excellent lumber, used in floors, and for the rich content of N-methyltryptamine and N,N-dimethyltryptamine in its tree bark. It is further considered, at least in Hawaii, to be an invasive species, once again, due to encroaching humanity. It is known in Chinese medicine as *hai hung tou* (red bean from the sea) and *hai yuk* (sea medicine). The plant's medicinal actions, associated with different plant parts, are said to include hepatoprotective and anti-inflammatory properties and to be due to flavonoids, phenolic acid derivatives, and flavonol glycoside (Lin et al. 2018). (a) *A. confusa* trees in Hawaii. (1/7/2005, Waikamoi trail, Maui, Hawaii: by Forest & Kim Starr, CC BY 4.0, via Starr Environmental, www.starrenvironmental.com/images/image/?q=24616126252.) (b) *A. confusa*, yellow flowers, buds, leaves, and immature seed pods. (6/8/2004, Lua Makika, Kahoolawe, Hawaii: by Forest & Kim Starr, CC BY 4.0, via Starr Environmental, www.starrenvironmental.com/images/image/?q=24083457534.)

AIDS may be delicately represented as a "social disease," and social excesses may certainly have contributed to the disease's development in a great number of cases, so it is ironical, or perhaps predictable according to the archaic Doctrine of Signatures, that a tree which is able to sustain such intense isolation as *V. nilotica* may have something to offer those whose hyper-social behavior caused them their crash. This equipoise in isolation by the botanical master of social distancing is reflected again in the photograph of a *V. nilotica* tree in Figure 1.6. Some of these trees grow in the desert, where they may seem to be the only ones for hundreds of kilometers.

V. nilotica is utilized by people indigenous to the areas in which it grows as a remedy for a variety of social diseases affecting principally the genitalia. In *in vitro* inhibition assays, aqueous, methanolic, and chloroformic extracts of the bark of the *babul*, the gum arabic tree of the great Middle Eastern deserts, significantly suppressed the growth of the Human Papillomavirus (HPV) and Herpes Simplex Virus Type 2 (HSV-2), both viruses responsible for genital infections causing morbidity, though in this setting, the extracts failed to significantly suppress HIV (Donalisio et al. 2018). Nonetheless, biosoluble gums from *V. nilotica* bark have been used in the manufacture of vaginal rings for carrying pharmaceutical drugs. These herb-drug combinations have exhibited clinical efficacy for inhibiting HIV (Han et al. 2007; Saxena et al. 2009).

(a)

(b) (c)

FIGURE 1.4 *Vachellia nilotica*, or *Acacia arabica*, the "gum arabic tree," is one of the most common, best known, and highly employed acacias in the Middle East. Its invasiveness is declaimed by epithets like "weed of national importance" in Australia and "federal noxious weed" in the United States. (a) *V. nilotica* branch with a flower, bud, and leaves. (2/26/2017, Kaikondrahalli Lake, Valliyamma Layout, Haralur, Karnataka 560103: by Ajit Ampalakkad, CC0 1.0, via iNaturalist, www.inaturalist.org/photos/157728699.) (b) *V. nilotica* seed pods. (4/20/2021, Roodeplaat, South Africa: by Hildegard Klein, CC BY 4.0, via iNaturalist, www.inaturalist.org/photos/122058659.) (c) *V. nilotica* tree. (4/20/2021, Roodeplaat, South Africa: by Hildegard Klein, CC BY 4.0, via iNaturalist, www.inaturalist.org/photos/122058781.)

(a) (b)

FIGURE 1.5 The *Acacia s.l.* tree, *Senegalia catechu*, the "black cutch," is the source of a wood extract, catechu, commonly used in folk medicine for sore throat and diarrhea. It is also employed as a rejuvenating tonic in Ayurveda, generically a *rasayana*. A powerful astringent, catechu from *S. catechu*, should not be confused with the related extract of the same name from the nut of another tree, the betel nut tree, *Areca catechu*. (a) *S. catechu* branch with leaves and white flowers. (India: by RealityImages, via Shutterstock, www.shutterstock.com/image-photo/twig-flowers-acacia-catechu-family-mimosaceae-1580936050.) (b) *S. catechu* tree. (Burma, Myanmar: by Sytilin Pavel, via Shutterstock, www.shutterstock.com/image-photo/acacia-catechu-khadira-burma-myanmar-176589191.)

FIGURE 1.6 Another view of *V. nilotica* illustrates its ostensible equanimity in radical isolation, a natural symbol for our quarantines during the novel coronavirus pandemic of 2019 onward. The tree is thus also meaningful for social diseases, of which one of the most extreme is acquired immune deficiency syndrome (AIDS). Photo: A lonely *Acacia* sp. in the Negev Desert. (Timna Park, Negev, Israel: by Jim Black, Pixabay License [https://pixabay.com/service/license/], via Pixabay, https://pixabay.com/photos/israel-negev-timna-park-rocks-4764747.)

While not as potent as the pharmaceutical antiviral Acyclovir, significant inhibition of both HSV and HIV was noted from a gel in which *S. catechu* was an important ingredient. The role of HSV as a possible factor in HIV development was highlighted (Mishra et al. 2018). A commercial product combining *S. catechu* extract with that of *Scutellaria baicalensis* radix was tested primarily for anti-inflammatory properties, though its antiviral effects were also emphasized (Bitto et al. 2014).

Proteins isolated from *A. confusa* seeds, i.e., acaconin (Lam and Ng 2010a) and acafusin (Lam and Ng 2010b), suppressed HIV activity *in vitro* through inhibition of HIV-1 reverse transcriptase. This enzyme catalyzes the conversion of the virion's single-stranded RNA into double-stranded DNA, the latter of which is far more suitable for occupying the host. Thus, HIV-1 reverse transcriptase is the core mechanism for the retrovirus's strategy as well as a key anti-AIDS pharmaceutical target.

Potential pharmacokinetic interactions between herbs used in African traditional medicine (ATM) and chemical retroviral drugs occur, so caution and consideration must be given to potential untoward interactions. In this context, *V. nilotica* was cited *vis-à-vis* pharmaceutical antiretrovirals (Müller and Kanfer 2011).

For influenza, extracts of the leaves of *V. nilotica* were highly and significantly virucidal, and they stopped the H9N2 avian flu virus from spreading in a live *in ovo* model (Ghoke et al. 2018). This was true even at very small doses. Also in eggs (i.e., *in ovo*), *V. nilotica* resin (i.e., gum arabic) inhibited the replication of red blood cells and their agglutination caused by influenza virus A (Green and Woolley 1947). In an investigation of seven samples with high tannin content—five natural products and two synthetic products—an *Acacia s.l.* product identified only as "wattle extract" had the second-highest activity *in vitro*, essentially equal to that of persimmon extract, which was minimally and insignificantly higher, when pitted against the enveloped, avian-originated human influenza A variants H3N2 and H5N3 (Ueda et al. 2013). Without a doubt, *Acacia s.l.* will provide important antiviral drugs, most notably in respiratory diseases for the time being; however, other significant large-scale encounters between humanity and viruses may lie ahead, posing new challenges for pharmacognosists and complex drug designers researching and developing this interface between plant, human virus, and human. *Acacia s.l.* will figure prominently in this noble planetary enterprise.

REFERENCES

Alatar, A., M.A. El-Sheikh, and J. Thomas. 2012. Vegetation analysis of Wadi Al-Jufair, a hyper-arid region in Najd, Saudi Arabia [retracted in 2012 in *Saudi J Biol Sci* 19(1): 43–54]. *Saudi J Biol Sci* 19(3): 357–8.

Arbab, A.H., M.K. Parvez, M.S. Al-Dosari, and A.J. Al-Rehaily. 2017. *In vitro* evaluation of novel antiviral activities of 60 medicinal plants extracts against hepatitis B virus. *Exp Ther Med* 14(1): 626–34.

Arbab, A.H., M.K. Parvez, M.S. Al-Dosari, A.J. Al-Rehaily, M. Al-Sohaibani, E.E. Zaroug, M.S. AlSaid, and S. Rafatullah. 2015. Hepatoprotective and antiviral efficacy of *Acacia mellifera* leaves fractions against hepatitis B virus. *Biomed Res Int* 2015: 929131.

Bhanuprakash, V., M. Hosamani, V. Balamurugan, P. Gandhale, Ram Naresh, D. Swarup, and R.K. Singh. 2008. *In vitro* antiviral activity of plant extracts on goatpox virus replication. *Indian J Exp Biol* 46(2): 120–7.

Bitto, A., F. Squadrito, N. Irrera, G. Pizzino, G. Pallio, A. Mecchio, F. Galfo, and D. Altavilla. 2014. Flavocoxid, a nutraceutical approach to blunt inflammatory conditions. *Mediators Inflamm* 2014: 790851.

Donalisio, M., V. Cagno, A. Civra, D. Gibellini, G. Musumeci, M. Rittà, M. Ghosh, and D. Lembo. 2018. The traditional use of *Vachellia nilotica* for sexually transmitted diseases is substantiated by the antiviral activity of its bark extract against sexually transmitted viruses. *J Ethnopharmacol* 213: 403–8.

Fetene, A., T. Hilker, K. Yeshitela, R. Prasse, W. Cohen, and Z. Yang. 2016. Detecting trends in landuse and landcover change of Nech Sar National Park, Ethiopia. *Environ Manage* 57(1): 137–47.

Ghoke, S.S., R. Sood, N. Kumar, A.K. Pateriya, S. Bhatia, A. Mishra, R. Dixit, V.K. Singh, D.N. Desai, D.D. Kulkarni, U. Dimri, and V.P. Singh. 2018. Evaluation of antiviral activity of *Ocimum sanctum* and *Acacia arabica* leaves extracts against H9N2 virus using embryonated chicken egg model. *BMC Complement Altern Med* 18(1): 174.

Green, R.H., and D.W. Woolley. 1947. Inhibition by certain polysaccharides of hemagglutination and of multiplication of influenza virus. *J Exp Med* 86(1): 55–64.

Han, Y.A., M. Singh, and B.B. Saxena. 2007. Development of vaginal rings for sustained release of nonhormonal contraceptives and anti-HIV agents. *Contraception* 76(2): 132–8.

Lam, S.K., and T.B. Ng. 2010a. Acaconin, a chitinase-like antifungal protein with cytotoxic and anti-HIV-1 reverse transcriptase activities from *Acacia confusa* seeds. *Acta Biochim Pol* 57(3): 299–304.

Lam, S.K., and T.B. Ng. 2010b. Acafusin, a dimeric antifungal protein from *Acacia confusa* seeds. *Protein Pept Lett* 17(7): 817–22.

Lee, J.C., W.C. Chen, S.F. Wu, C.K. Tseng, C.Y. Chiou, F.R. Chang, S.H. Hsu, and Y.C. Wu. 2011. Anti-hepatitis C virus activity of *Acacia confusa* extract via suppressing cyclooxygenase-2. *Antiviral Res* 89(1): 35–42.

Lin, H.Y., T.C. Chang, and S.T. Chang. 2018. A review of antioxidant and pharmacological properties of phenolic compounds in *Acacia confusa*. *J Tradit Complement Med* 8(4): 443–50.

Mishra, N.N., A. Kesharwani, A. Agarwal, S.K. Polachira, R. Nair, and S.K. Gupta. 2018. Herbal gel formulation developed for anti-human immunodeficiency virus (HIV)-1 activity also inhibits *in vitro* HSV-2 infection. *Viruses* 10(11): 580.

Müller, A.C., and I. Kanfer. 2011. Potential pharmacokinetic interactions between antiretrovirals and medicinal plants used as complementary and African traditional medicines. *Biopharm Drug Dispos* 32(8): 458–70.

Panya, A., P. Yongpitakwattana, P. Budchart, N. Sawasdee, S. Krobthong, A. Paemanee, S. Roytrakul, S. Rattanabunyong, K. Choowongkomon, and P.T. Yenchitsomanus. 2019. Novel bioactive peptides demonstrating anti-dengue virus activity isolated from the Asian medicinal plant *Acacia Catechu*. *Chem Biol Drug Des* 93(2): 100–9.

Raheel, R., M. Ashraf, S. Ejaz, A. Javeed, and I. Altaf. 2013. Assessment of the cytotoxic and anti-viral potential of aqueous extracts from different parts of *Acacia nilotica* (Linn) Delile against Peste des petits ruminants virus. *Environ Toxicol Pharmacol* 35(1): 72–81.

Rehman, S., U.A. Ashfaq, S. Riaz, T. Javed, and S. Riazuddin. 2011. Antiviral activity of *Acacia nilotica* against hepatitis C virus in liver infected cells. *Virol J* 8: 220.

Saxena, B.B., Y.A. Han, D. Fu, P. Rathnam, M. Singh, J. Laurence, and S. Lerner. 2009. Sustained release of microbicides by newly engineered vaginal rings. *AIDS* 23(8): 917–22.

Ueda, K., R. Kawabata, T. Irie, Y. Nakai, Y. Tohya, and T. Sakaguchi. 2013. Inactivation of pathogenic viruses by plant-derived tannins: Strong effects of extracts from persimmon (*Diospyros kaki*) on a broad range of viruses. *PLoS One* 8(1): e55343.

Zarei, M., J. Asgarpanah, and P. Ziarati. 2015. Chemical composition profile of wild *Acacia oerfota* (Forssk) Schweinf seed growing in the South of Iran. *Orient J Chem* 31(4): 2311–18.

2 Antibacterial Effects of Acacias

Bacteria are not viruses, so what's the point in the midst of the pandemic? There are two reasons.

First, when a virus occupies its host, the host is weakened and consequently more susceptible to "secondary" bacterial infection. Though the virus incapacitates its host, a bacterium often deals the final blow.

The second major reason is that bacteria can also deliver the entire crippling and lethal process even without the virus's assistance. The 1918 flu, it is true, was caused by a virus, but in cholera, tuberculosis, gonorrhea, and the Black Plague, bacteria are and were the agents of contagion. The immediacy of the need for novel antibacterials from Nature is presented by antibiotic-resistant bacteria. These prokaryotic cells evolved rapidly beyond the antibiotic threat to their existence and found ways to resist it. Conceivably, such bacterial resistance could also be developed for natural products extracted from the leaves, bark, flowers, fruits, and roots of plants, but the complexity of these crude drugs makes the work of such impromptu bacterial evolution more difficult and complex.

Botanically sourced antibiotics are always of interest to the pharmaceutical industry, if not for their direct incorporation into drugs, then as "leads" for the development of new pure synthetic drugs. Complex botanical drugs, such as those directly derived from biomass, have a certain advantage for therapeutics, namely that the intricacy of synergistic interactions presents a logistical obstacle to invasion and occupation. As bacteria evolve to overcome previously effective single-chemical agents, complex drugs may become more than just an ecological alternative drug in an increasing number of clinical settings.

One very simple example of a complex agent is the combination of a therapeutic carrier, such as one that supports immunity and facile absorption, with an active agent. In the development of such carriers, the *Acacia s.l.* genus has played a pivotal role, primarily via the gums exuded by its trees, and most specifically by "gum arabic," officially derived from *Senegalia senegal* but also sometimes *ad hoc* from other *Acacia s.l.* species (see also Chapter 4, Anti-Inflammatory Influence of Acacias). Metallic silver, in nano dosage, is an example of a naturally occurring, maximally simple antibiotic, employed even today in modern medicine for specific applications such as sterile, bacteria-resistant dressings and bandages for severe burns. Silver nanoparticles have been repeatedly used as therapeutic drugs in an immuno-supportive and absorption-friendly matrix derived from acacia trees (Aadil et al. 2018; Bajpai and Kumari 2015; Juby et al. 2012; Rao et al. 2019), for example as an antibacterial treatment against *Pseudomonas aeruginosa* (Ansari et al. 2014), as have zinc oxide (Bajpai et al. 2016; Chopra et al. 2015; Raguvaran et al. 2017; Thakur et al. 2013) and classical antibacterial antibiotics (Bartzoka et al. 2018; Daniels 1949; Eltayeb et al. 2004).

One recent study combined newly synthesized silver nanoparticles with gums derived from *Vachellia nilotica* (*kikar*), *Senegalia modesta* (*phulai*), *Prunus armeniaca* (apricot), *Prunus domestica* (plums), *Prunus persica* (peaches), or *Salmalia malabarica* (silk cotton tree), in which efficacy was enhanced and toxicity contained (Jamila et al. 2020). In all these settings, gum arabic is both immunopotent and directly antibacterial (Baien et al. 2020), with the combinations most likely embodying therapeutic synergy. Such gum arabic matrices find employment both for oral drugs, for example in the gel capsules that contain them, and in dressings for wounds (Bhatnagar et al. 2013) or burns. In green chemistry applications, the *V. nilotica* extracts act as reducing and stabilizing agents for the creation of stable silver nanoparticles . . ., i.e., the extracts "create" the nanoparticles (Mohammed et al. 2018)!

As well, *V. nilotica* extracts have been used in antibiotic formulae combined with other natural products (Chandra Shekar et al. 2016). Clark et al. (1993) found that gum arabic has a lot of antibiotic potential in dentistry for treating gingivitis. This means that it can prevent tooth loss.

DOI: 10.1201/9780429440946-2

An extract of *Senegalia catechu* was also used with other herbs in a complex botanical gel to make a powerful topical antibiotic whose activity was higher than that of the standard drug Cloxacillin (Dashtdar et al. 2016) and an acne treatment whose *in vitro* activity against *Propionibacterium acnes*, *Staphylococcus epidermidis*, and *Staphylococcus aureus* was similar to that of clindamycin (Nipanikar et al. 2017). In a large-scale modern screening of 239 Chinese medicinal plants against multidrug-resistant *Staphylococcus aureus* (MRSA), *S. catechu* was among the best, encompassing both the highest antibiotic activity and the lowest cytotoxicity (Kim et al. 2020). Catechu was also formerly and famously the subject of clinical trials for the treatment of lepromatous leprosy (Ojha et al. 1969). It is also one of a handful of plants out of 38 used in Thailand for treatment of diarrhea that significantly inhibited (MIC) and killed several strains of enterohemorrhagic *E. coli* (Voravuthikunchai et al. 2004), at least in part by modifying the cell surface hydrophobicity of the bacteria (Voravuthikunchai and Limsuwan 2006).

A composition of catechu and an extract of *Scutellaria baicalensis* were administered to 20 male and female dogs after they had been induced into an artificial state of periodontitis by ligature. The combination had an excellent effect on the dogs' periodontal health, with statistically significant reductions in gingivitis, pocket depth, loss of attachment, and gum bleeding (Yimam et al. 2019).

In another case, Escárcega-González et al. (2018) used elements from *Vachellia rigidula* as a "reducing and capping agent" to make silver nanoparticles in a "green" system.

FIGURE 2.1 *Scutellaria baicalensis* (Baikal skullcap). The root is the part used in classical Chinese medicine. Photo: *S. baicalensis* Georgi flowers and leaves. (Kawaguchi Green Center, Kawaguchi, Saitama Prefecture, Japan: by Tanaka Juuyoh, CC BY 2.0, via Flickr, www.flickr.com/photos/tanaka_juuyoh/2718717267.)

(b)

(a) (c)

FIGURE 2.2 The gum of the *Senegalia catechu*, black cutch tree, produced by concentrating the aqueous extract of the heartwood, is used as an ingredient in *quid*, a mixture, usually of cutch heartwood extract with tobacco and/or the areca nut (*Areca catechu*—not to be confused with *Acacia catechu*) and betel leaf (*Piper betle*). Cutch extract, also known as *kher*, or *catechu*, is a powerful astringent that can be used to treat sore throats and diarrhea. The name *catechu* gave rise to our modern *catechol*, catecholamine, etc., in tribute to the alkaloids in this class supposedly first discovered in the heartwood. Other common names for cutch extract are summarized in Table 2.1 (after Quattrocchi 2000). (a) *Acacia catechu* Willd. (= *S. catechu*) branch with leaves and flowers, mature seed pod, and seeds; botanical drawing. (In F. E. Köhler, *Köhler's Medizinal-Pflanzen in naturgetreuen Abbildungen mit kurz erläuterndem Texte* [Gera-Untermhaus: Fr. Eugen Köhler, 1883–1914], vol. 2, Table 116, via www.plantillustrations.org, public domain, www.plantillustrations.org/illustration.php?id_illustration=31383.) (b) *S. catechu* trunk. Catechu is extracted from the tree's heartwood. (2/4/2021, Lweta, Uttarakhand, India: by K. Ramnarayan, CC BY 4.0, via iNaturalist, www.inaturalist.org/photos/112037696.) (c) Catechu pieces. (By Simmax, via Shutterstock, www.shutterstock.com/image-photo/acacia-catechu-on-white-background-207355810.)

TABLE 2.1
Common Names for Extract of *Senegalia catechu* Heartwood

Country/Language	Name(s)
English	Black cutch, wadali-gumtree, Jerusalem thorn, cutch, catechu
India	Khadira, khayer, khoiru, kanchu, khair, karangalli, kachu, kaderi, kagh, lal-khair, shemi, karan-galli, koir, sandra
China	Hai er cha, erh ch'a, wu tieh nl
Brazil	Catechu, cato, cato-de-pego, terra-japonica
Kenya and Tanzania	Mgenda, mgunga
Tibetan	Seng-ldeng, gsom seng-ldeng, tsan-dan sang-ldeng
Burma	Nya, sha, shaji, tun-sa-se

Source: After Quattrocchi (2000).

FIGURE 2.3 *Senegalia modesta, Phulai* (Pakistan), *Phalāhī* (India) are also found in Afghanistan and Africa. Traditional uses include leprosy, wounds, dysentery, and venereal diseases (Sarwar 2016). *Acacia modesta* Wall. (= *S. modesta*) branch with flowers and leaves; botanical drawing. (By Vishnupersaud, in N. Wallich, *Plantae Asiaticae Rariores* [London: Treuttel and Würtz, 1830–32], via www.plantillustrations.org, public domain, www.plantillustrations.org/illustration.php?id_illustration=10783.)

(a) (b)

FIGURE 2.4 Blackbrush acacia (*V. rigidula*), also known as *chaparro prieto* for its short stature, is a small tree or shrub native to the New World, from Texas south into Mexico. Disputed analytical chemical studies have found dimethyltryptamine (DMT) in its parts as well as nicotine and even mescaline in very small amounts. Similar psychoactive profiles were found by the same research group in a closely related *Acacia s.l.* species, *Senegalia berlandieri*, also endemic to the same geographical area. Both trees were studied to differentiate them from each other because of their association with a bovine locomotor ataxia, "limber leg," attributable to *S. berlandieri*, though not to *V. rigidula*. (a) *V. rigidula*, leaves and flowers. (3/8/2018, Calle José Camilo Cela 108–144, Guadalupe, NL, México: by Adriana Nelly Correa Sandoval, CC BY 4.0, via iNaturalist, www.inaturalist.org/photos/13999300.) (b) Bush-shaped *V. rigidula*. (3/8/2018, Calle José Camilo Cela 108–144, Guadalupe, NL, México: by Adriana Nelly Correa Sandoval, CC BY 4.0, via iNaturalist, www.inaturalist.org/photos/13999287.)

(a) (b) (c)

FIGURE 2.5 *Senegalia berlandieri*, berlandier acacia, *guajillo*, *huajillo*, restricted to arid regions, is said, when rainfall does occur in just the right amount, to produce a clear and practically colorless "best honey on the planet," according to one senior beekeeper and honey expert (www.grampashoney.com). The leaves contain alkaloids, including, in the highest amounts, N-methylphenethylamine, tyramine, and phenethylamine, with possible trace amounts of amphetamine, nicotine, and mescaline (Clement et al. 2007). (a) *S. berlandieri*, leaves and white flowers. (4/6/2018, Val Verde County, TX: by Alison Northup, CC BY 4.0, via iNaturalist, www.inaturalist.org/photos/15000534.) (b) *S. berlandieri* leaves and mature seed pods. Leaves may contain psychoactive alkaloids. (6/26/2020, Kickapoo Cavern State Park, Brackettville, TX: by Cody Stricker, CC BY 4.0, via iNaturalist, www.inaturalist.org/photos/81423304.) (c) *S. berlandieri* bush in Texas. (4/6/2018, 9685 US-90, Del Rio, TX 78840: by Alison Northup, CC BY 4.0, via iNaturalist, www.inaturalist.org/photos/38061665.)

Farag et al. (2015) selected 25 different medicinal plants known in Egyptian folk medicine to have strong "antiseptic relevance" and tested their extracts against the potato, eggplant, and tomato pathogen *Ralstonia solanacearum* that causes bacterial wilt. The extracts with the highest activity against the wilt were, in first place, *Vachellia nilotica*, and in second place, the pomegranate, *Punica granatum*. Epicatechin, gallic acid, and 3,5-dihydroxy-4-methoxybenzoic acid (4-OMGA), which is a metabolite of gallic acid, are all antibiotic compounds found in both species. 4-OMGA is the most powerful antibiotic compound found in both species.

Extracts of *V. nilotica* have been extensively studied and employed for their antibacterial properties. Gupta and Gupta (2015) conducted a clinical trial with 90 patients at high risk for dental caries over a thirty-day period. The first group rinsed each morning with a mouthwash consisting of a 50% *V. nilotica* extract (VNE), the second with a usual dose of chlorhexidine, and the third with saline. The VNE was comparable—and actually a little superior—to the chlorhexidine, though in another study, two *V. nilotica* products, gel and powder, were shown to have an effect approximately equal to each other and to 1% chlorhexidine (Pradeep et al. 2012). Similarly, in a study of extracts from three plants, *V. nilotica*, *Tamarix aphylla*, and *Melia azadirachta*, against bacteria causing biofilm resulting in dental caries and orthodontic disease from human caries isolates, *V. nilotica* was the most effective. It not only radically inhibited the growth of the pathogenic bacteria but was also efficacious at cleaning existing biofilms (Khalid et al. 2017).

In an ethnographic survey conducted in Zimbabwe, *V. nilotica* was found to be the most commonly used herb for treating venereal infections. Extracts from methanol or acetone were more active against a panel of Gram-positive and Gram-negative bacteria in an *in vitro* trial than the equally active aqueous extracts (Kambizi and Afolayan 2001).

Even with the ongoing development of the most sophisticated new antibiotic drugs, bacteria adapt, manifesting in the vital clinical problem of multidrug resistant (MDR) bacterial strains. The strong resistance of such strains to naturally antibiotic plant extracts from the Ayurvedic staple *Terminalia arjuna* and the great tree *Eucalyptus globulus* demonstrates that this phenomenon is not limited to pharmaceutical antibiotics. These MDR strains were inhibited by extracts of common cloves, *Syzygium aromaticum*, and cinnamon, *Cinnamomum zeylanicum*, but most potently by *V. nilotica* (Khan et al. 2009).

Mahmood et al. (2012) studied the methanolic extract of *V. nilotica* leaves and two related non-*Acacia s.l.* Mimosaceae species against a small selection of Gram-positive and Gram-negative bacterial species. Although there was little suppression of the Gram-positive *Bacillus subtilis*, they did strongly attenuate the growth of Gram-negative *Escherichia coli*, *Pseudomonas aeruginosa*, and *Klebsiella pneumonia*. Fruit pericarps of *V. nilotica* also showed "massively" good activity against *E. coli in vitro* (Mazzio et al. 2016).

Mishra et al. (2016) found that *Escherichia, Klebsiella, Salmonella, Pseudomonas*, and *Staphylococcus* bacteria were killed by methyl gallate, a gallic acid derivative that was isolated from *V. nilotica* leaf extract.

Both *V. nilotica* and *S. catechu* extracts among 54 extracts tested suppressed multi-drug resistant *Salmonella typhi* growth in a screening assay, though less strongly than *Punica granatum* and a number of other Ayurvedic medicinal herbs or their combinations, *viz. Myristica fragrans*, *Aegle marmelos*, *Salmalia malabarica*, *Holarrhena antidysenterica*, *Terminalia arjuna*, and *Triphal* (mixture of *Emblica officinalis*, *Terminalia chebula*, and *Terminalia bellirica*) (Rani and Khullar 2004). Also, *V. nilotica* significantly inhibited bacterial biofilm growth indicative of otitis media, as did ciprofloxacin, acting against *Pseudomonas aeruginosa*, *Staphylococcus haemolyticus*, and *Staphylococcus hominis* (Rehman et al. 2016).

A *V. nilotica* extract and a compound isolated from the extract, pyrogallol, respectively inhibited growth of a multidrug resistant strain of *Helicobacter pyloris* at 20 μg/mL and 10 μg/mL (Revathi et al. 2018). A hot aqueous extract of *V. nilotica* leaves significantly inhibited *Pseudomonas aeruginosa, Escherichia coli, Bacillus cereus, Staphylococcus aureus,* and *Streptococcus uberis in vitro* (Sharma et al. 2014). In 3- to 4-year-old male Baladi goats (n = 25), inclusion of *V. nilotica* leaves in their diets inhibited the degree of total bacteria and the pathogenic *Clostridium perfringens* in their ruminal juice and feces. The effect was attributed to a 34% total tannin content in the leaves (Sotohy et al. 1997).

In a study of the antimicrobial effects of different "chewing sticks" used in Pakistan, aqueous extracts of *V. nilotica* sticks showed good antibacterial activity against *Streptococcus fecalis* at 50% concentration (Almas 2001). Extracts of different parts of *V. nilotica* were more potent than the antibiotic metronidazole and nearly as effective as tetracycline, though less effective than amoxicillin or clarithromycin, against *Helicobacter pylori* (Amin et al. 2013). A strong inhibition of the enzyme urease, which catalyzes the hydrolysis of urea to ammonia and carbon dioxide and whose action is contributory to bacterial urinary tract infections, was noted in *V. nilotica* parts extracted in methanol, water, or cow's urine (Bai et al. 2015). In another Indian study, ethyl acetate proved the solvent of maximum efficacy for extracting the anti-cariogenic principal from *V. nilotica*, particularly potent against the tooth-rotting *Streptococcus mutans* and *Lactobacillus casei* (Barad et al. 2014). Aqueous extracts of *V. nilotica* quelled *Salmonella typhi, Escherichia coli, Pseudomonas aeruginosa, Staphylococcus aureus*, and *Bacillus cereus* at a minimum inhibitory concentration (MIC) averaging 17.5 μg/mL, while methanolic extracts were somewhat less effective, averaging 75 μg/mL. Ethanolic extracts were not tested in this study (Dabur et al. 2007).

Of course, traditional usage almost always employs aqueous extraction, and results are often contradictory as to which solvent used for extraction provides superior results. Methanol is a great favorite for bench research, though it is rarely used traditionally. Kenyans have used *V. nilotica* for a long time to treat lower respiratory tract infections. Kariuki and Njoroge (2011) found that methanol extracts of the tree's roots and bark were effective against *Staphylococcus aureus, Streptococcus pneumoniae*, and *Escherichia coli.*

FIGURE 2.6 Egyptian goat. (1/18/2012: by Essam Saad, CC BY 2.0, via Flickr, www.flickr.com/photos/essamsaad/8443265085.)

Extracts from *V. nilotica* fruit pods, or pericarps, also possess antibiotic activity and should not be overlooked, neither for pharmaceutical-driven research nor in ethnomedical usage. Fruit pericarps are usually thrown away, so they are an attractive, low-cost source of pharmacologically active biomass that does not harm the tree much when it is collected (Karim and Azlan 2012).

The extracts of *V. nilotica*, *Caesalpinia bonducella*, and *Gardenia gummifera* showed the most antibacterial activity against *Escherichia coli, Staphylococcus aureus, Klebsiella pneumoniae, Proteus vulgaris, Salmonella typhi, Shigella flexneri, Salmonella paratyphi, Salmonella typhimurium, Pseudomonas aeruginosa*, and *Enterobacter aerogenes* by disc diffusion method (Tambekar et al. 2009).

Niloticane, a novel cassane diterpene, was isolated from the ethyl acetate bark extract of *V. nilotica* subsp. *kraussiana*. At 4 and 8 μg/mL MIC, respectively, the extract and niloticane were very active against Gram-positive bacteria *Bacillus subtilis* and *Staphylococcus aureus*. Milder activity against Gram-negative *Klebsiella pneumoniae* and *E. coli* was shown at 16 and 33 μg/mL, respectively (Eldeen et al. 2010).

V. nilotica, from its famous gum arabic (though it is not the only *Acacia s.l.* species to make this claim) to its leaves, is difficult to follow for other species in this super *Acacia s.l.* "genus," but many other species within *Acacia s.l.* exhibit pharmacological properties of note.

One grouping to consider is the representation of *Acacia s.l.* in Australia, which is usually of the remaining genus *Acacia*. Most of the *Acacia s.l.* as a whole are in Australia.

In a study for antibacterial efficacy among 8 plants used medicinally by Aboriginal people in New South Wales, *Acacia implexa* and *Acacia falcata* showed good and significant inhibition on discs of multidrug resistant strains of *Staphylococcus aureus*, with minimum inhibitory concentrations from 7.81 to 1000 μg/mL, leading to interest for their use as botanical drugs for topical use in skin disorders caused by *S. aureus* (Akter et al. 2016).

(a) (b)

FIGURE 2.7 *Acacia implexa*, "lightwood" or "hickory wattle," grows to be 5 to 15 m high and up to 10 m wide. Its wood is prized for furniture. It is considered a fast-growing, shade-intolerant species. (a) *A. implexa* Benth., branch with leaves and cream-colored flowers. (2/16/2011, Black Mountain, Canberra, ACT, Australia: by Donald Hobern, CC BY 2.0, via Flickr, www.flickr.com/photos/dhobern/5449454544.) (b) *A. implexa* tree. (5/9/2020, Stellenbosch, Western Cape, South Africa: by Dave Richardson, CC BY 4.0, via iNaturalist, www.inaturalist.org/photos/71967079.)

(a) (b)

FIGURE 2.8 *Acacia falcata*, sickle wattle, also known as "Sally," is a large shrub or small tree, reaching a typical height of 2–5 m. The bark is very rich in tannins, and it has been used by Aboriginal people in Australia for an external treatment of skin diseases and to stupefy fish. (a) *A. falcata*, leaves and white flowers. (4/30/2018, 21 Retreat Ct, Bunya QLD 4055, Australia: by Sam Stainsby, CC BY 4.0, via iNaturalist, www.inaturalist.org/photos/16986449.) (b) *A. falcata*, leaves and unripe seed pods. (4/30/2018, 21 Retreat Ct, Bunya QLD 4055, Australia: by Sam Stainsby, CC BY 4.0, via iNaturalist, www.inaturalist.org/photos/17059787.)

(c) (d)

FIGURE 2.8 (Continued) (c) *A. falcata*, tree trunk and bark. (4/30/2018, 21 Retreat Ct, Bunya QLD 4055, Australia: by Sam Stainsby, CC BY 4.0, via iNaturalist, www.inaturalist.org/photos/17058935.) (d) *A. falcata*, habitat. (4/30/2018, 21 Retreat Ct, Bunya QLD 4055, Australia: by Sam Stainsby, CC BY 4.0, via iNaturalist, www.inaturalist.org/photos/17059394.)

FIGURE 2.9 *Jalmenus evagoras*, Common Imperial Blue, depends on *A. mearnsii*. (4/11/2022, Yatte Yattah NSW 2539, Australia: by Christopher Brandis, CC0 1.0, via iNaturalist, www.inaturalist.org/photos/189357873.)

Another great and beloved Australian acacia tree is the "black wattle," *Acacia mearnsii*, named for the American naturalist Edgar Alexander Mearns (1895–1916) (Quattrocchi 2000) and the subject of a fine book-length treatise (Sherry 1968). The tree is both considered an integral part of Australia's ecosystem, hosting key pollinators such as *Jalmenus evagoras*, fixing the soil following brush fires, and providing its fixed nitrogen to other valuable plant species, and simultaneously regarded as the most viciously invasive plant on the planet (www.iucngisd.org). A tannin-rich extract from its bark was shown to capably inhibit growth of the common intestinal bacteria *Escherichia coli*, *Klebsiella pneumoniae*, *Proteus vulgaris* and *Serratia marcescens* as well as the so-called "bad bacteria" of the intestine, genera *Clostridium* and *Bacteroides* (Ogawa and Yazaki 2018). Acetone extract of tree parts of *A. mearnsii*, naturalized in South Africa, was also effective against a panel of multidrug resistant enteric pathogenic bacteria responsible for diarrhea in Africa, another debarkation point of the globally "invasive" black wattle (Bisi-Johnson et al. 2017).

Traditionally used to treat digestive disorders and diarrhea, *A. mearnsii* extracts have demonstrated important antibiotic activity against Gram-negative pathogenic intestinal bacteria of the genus *Shigella* (Olajuyigbe and Afolayan 2012a), and synergized successfully in combination with conventional antibiotics (Olajuyigbe and Afolayan 2012c). Further, a high margin of safety, i.e., a wide therapeutic window, was established for the use of the bactericidal *A. mearnsii* heartwood and bark extracts for treating a wide range of bacterial infections clinically (Olajuyigbe and Afolayan 2012b). Essential oils produced by hydrodistillation of *A. mearnsii* (a.k.a. *Acacia mollissima*) and *Acacia cyclops* were found to be active against a panel of Gram-positive and Gram-negative bacteria, with the greatest activity by Tunisian-grown *A. mearnsii* against *Pseudomonas aeruginosa* at a minimum inhibitory concentration of 0.31 mg/mL (310 μg/mL) (Jelassi et al. 2017). Curiously, *A. mearnsii* root systems seem capable of responding to changes in carbon dioxide in the environment due to climate change and can accordingly modulate their production and secretion of antibiotic compounds (Wu and Yu 2019).

FIGURE 2.10 *Acacia mearnsii* flowers. Black wattle is a rapidly growing tree that is essential to the Australian ecosystem, and an invasive species worldwide. In Australia, its high-tannin bark is prized for tanning leather and as medicine, and its wood is used as lumber. The heartwood contains a flavan-3.4-diol, leuco-fisetinidin, an anthocyanidin congener in a condensed tannin complex. (12/24/2005, Polipoli, Maui, Hawaii: by Forest & Kim Starr, CC BY 4.0, via Starr Environmental, www.starrenvironmental.com/images/image/?q=24222188124.)

Seed extracts of three other Australian *Acacia* species, *Acacia cyclops*, *A. microbotrya*, and *A. victoriae*, were tested for antibacterial activity against common food-borne pathogens, *viz. Bacillus cereus*, *Escherichia coli*, *Salmonella typhimurium*, and *Staphylococcus aureus.* All the seeds showed bactericidal activity, with the strongest deriving from *A. cyclops*, underscoring the potential of the seeds of these acacias for food preservation and therapeutic nutrition (Chong et al. 2019).

(a) (b) (c)

FIGURE 2.11 *Acacia cyclops*, also known as coastal wattle, cyclops wattle, or red-eyed wattle, is a common coastal species that is considered highly invasive. It has turned up in South Africa and California as well as in unwanted settings within its native Australia. (a) *A. cyclops*, mature seed pod. (12/7/2009: by Sydney Oats, CC BY 2.0, via Flickr, www.flickr.com/photos/57768042@N00/4167478617.) (b) *A. cyclops*, blossoming tree. (4/6/2009: by Sydney Oats, CC BY 2.0, via Flickr, www.flickr.com/photos/57768042@N00/3417246586.) (c) *A. cyclops* flower. (5/6/2020, Stellenbosch, South Africa: by Dave Richardson, via iNaturalist, CC BY 4.0, www.inaturalist.org/photos/71429342.)

(a) (b)

FIGURE 2.12 *Acacia microbotrya*, also known as manna wattle, or gum wattle, is a fast-growing, salt-tolerant bushy shrub or good-sized tree, 2–7 m in height and up to 5 m wide. It is popular in native gardens as an ornamental or as a windbreak, and also for land rehabilitation. (a) *A. microbotrya* seed pods. Seed extract is antibacterial. (6/20/2007: by Jean and Fred Hort, CC BY 2.0, via Flickr, www.flickr.com/photos/jean_hort/51958604312/.) (b) *A. microbotrya* trunk, with gum hardening on the bark. (3/23/2022: by Jean and Fred Hort, CC BY 2.0, via Flickr, www.flickr.com/photos/jean_hort/51958597772/.)

(c)

FIGURE 2.12 (Continued) (c) *A. microbotrya* tree. (3/23/2022: by Jean and Fred Hort, CC BY 2.0, via Flickr, www.flickr.com/photos/jean_hort/51959881704/.)

(a) (b) (c)

FIGURE 2.13 *Acacia victoriae*, gundabluey, or bardi bush, is a small shrub-like tree. Its seeds provide food for mammals and birds, for whom it is also a refuge from the forces of the Australian arid zones and clay soils in which it is found. (a) *A. victoriae* tree in its habitat in Australia. (9/20/2020, Angus Smith Drive, Mount Stuart, Queensland, Australia: by Elawrey, CC BY 4.0, via iNaturalist, www.inaturalist.org/photos/98807608.) (b) *A. victoriae*, leaves and seed pods. (12/24/2018, Marino Conservation Park, SA, Australia: by Geoffrey Cox, CC BY 4.0, via iNaturalist, www.inaturalist.org/photos/29680410.) (c) *A. victoriae* seeds. (Photo via ScienceImage, CSIRO, CC BY 3.0, www.scienceimage.csiro.au/image/4021/seeds-of-the-elegant-wattle-acacia-victoriae.)

Leaf and bark extracts of another broad and bushy Australian species, *Acacia ligulata*, were active against *Streptococcus pyogenes* with a minimum inhibitory concentration of 62.5 μg/mL (Jæger et al. 2018). *A. ligulata* is considered to be psychoactive (Saroya and Singh 2020).

An extract of *Acacia pycnantha*, the "golden wattle" and the national emblem of Australia (Hitchcock 2012), was tested against bacteria, but at the dose tested, 32 μg/mL, was inactive

(a)

(b)

FIGURE 2.14 *Acacia ligulata*, sand dune wattle, the small *coobah*, is a dense "umbrella bush," its Latin name *ligulata* referring to the tongue-like shape of its blue-green, leaf-like phyllodes. The ashes were—and still are—mixed by Indigenous Australians with the dried leaves of the nicotine-containing *Duboisia hopwoodii* to generate visions. *A. ligulata* diaspores containing the seeds are transported by both ants and birds, most notably the "red wattle bird" (*Anthochaera carunculata*). This strategy has made this wattle the most widely distributed in Australia. (a) *A. ligulata* bush growing on a sandy dune. (1/10/2021, Colignan VIC 3494, Australia: by Dylan Butcher, CC BY 4.0, via iNaturalist, www.inaturalist.org/photos/109806173.) (b) *A. ligulata* leaves, close-up. (1/10/2021, Colignan VIC 3494, Australia: by Dylan Butcher, CC BY 4.0, via iNaturalist, www.inaturalist.org/photos/109806183.)

FIGURE 2.15 *Anthochaera carunculata*, the red wattlebird—essential to the success of *Acacia ligulata*. (8/22/2013: by Jean and Fred Hort, CC BY 2.0, via Flickr, www.flickr.com/photos/jean_hort/9619521961.)

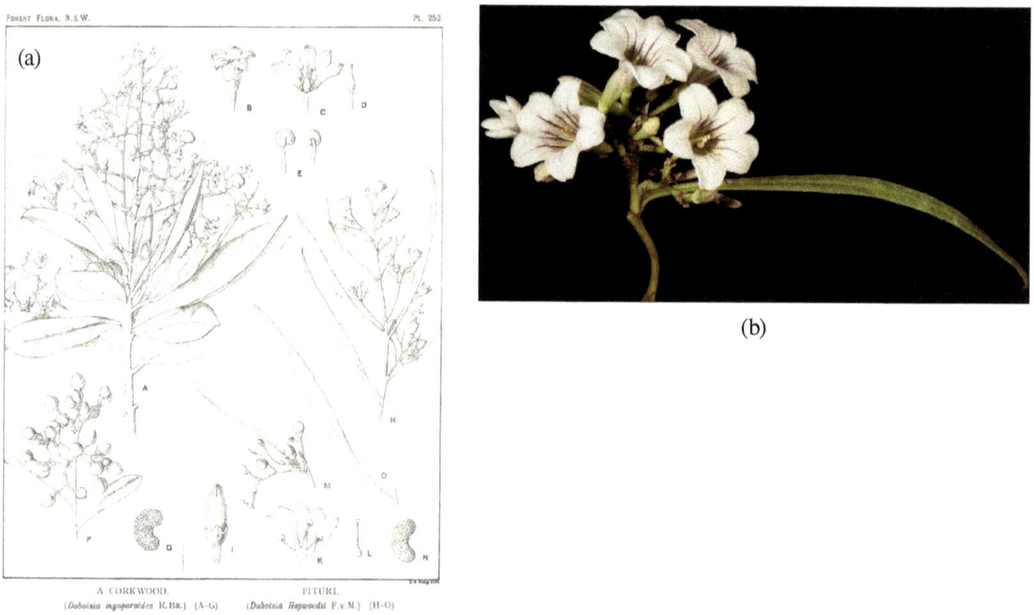

FIGURE 2.16 *Duboisia hopwoodii*. The dried leaves of the shrub are combined with the ashes of *Acacia ligulata* to produce *pituri*, a mixture with reported entheogenic properties. (a) *D. myoporoides* R.Br. and *D. hopwoodii* F.v.M., botanical drawing. (In J. H. Maiden, *The Forest Flora of New South Wales*, Vol. 7 [Sydney: W. A. Gullick, 1917], via www.plantillustrations.org, public domain, www.plantillustrations.org/illustration.php?id_illustration=63322.) (b) *D. hopwoodii*, flowers and leaf, close-up. (10/20/2013: by Kevin Thiele, CC BY 2.0, via Flickr, www.flickr.com/photos/66951228@N07/15435017332/.)

(a) (b)

FIGURE 2.17 *Acacia pycnantha*, golden wattle, can grow as high as 8 m. Like many other *Acacia* species, it has phyllodes (flattened leaf stalks) instead of true leaves. Its bark is said to contain more tannin than any other species of *Acacia*. Golden wattle is the official symbol of Australia. (a) *A. pycnantha*, flowers and leaves. (6/16/2014: by Candiru, public domain, via Flickr, www.flickr.com/photos/28061028@N07/14435315264/.) (b) *A. pycnantha*, bark and young shoots. (12/19/2020, Hallett Cove, SA, Australia: by Geoffrey Cox, CC BY 4.0, via iNaturalist, www.inaturalist.org/photos/108381507.)

in one study (Hendra et al. 2019), though its antibacterial *in vitro* efficacy at higher doses against *Klebsiella oxytoca*, *Staphylococcus aureus*, and *Klebsiella pneumoniae* had been shown (Mahmoud et al. 2016).

An extract of another Australian species, *Acacia decurrens*, was combined with an extract of a non-*Acacia* Australian species, *Eremophila glabra*, the essential oil of another Australian species, *Backhousia citriodora* (lemon myrtle), and some purified terpenoids, along with the antibiotic virginiamycin, and the mixture was administered to broiler chickens over 7 weeks for possible utility in controlling the pathogen *Campylobacter jejuni*. The *E. glabra* was determined to exert a negative effect on the putative anti-*Campylobacter* activity, and the overall mixture failed to score significance. Therefore, not much can be extracted from this study other than possible heuristic curiosity regarding *A. decurrens* and a healthy appreciation of the inherent difficulties in studying a preparation of such complexity (Kurekci et al. 2014).

Pennacchio et al. (2005) recounted their studies of four Australian plants known to ethnobotanists for their medicinal properties. Among them were two *Acacia* species, *A. auriculiformis* and *A. bivenosa*, which exhibited meaningful antibiotic activity. *A. auriculiformis* also was active against methicillin-resistant *Staphylococcus aureus* (MRSA) at doses less than 500 µg/disc (Chew et al. 2011), whereas a mixture of partially purified novel *A. auriculiformis* saponins, *acaciaside a* and *b*, inhibited growth of *Bacillus megaterium*, *Salmonella typhimurium*, and *Pseudomonas aeruginosa* at 700 µg/mL (Mandal et al. 2005). An aqueous extract of seed pods from *A. auriculiformis* was utilized as a reducing and capping agent in the manufacture of silver nanoparticles (Nalawade et al. 2014).

In a study of five traditional Australian medicinal plants, *Acacia kempeana* leaf extract exhibited a bacteriostatic effect against vancomycin-resistant enterococci (Palombo and Semple 2002).

FIGURE 2.18 *Acacia decurrens*. Not to be confused with the black wattle, *Acacia mearnsii*, which it somewhat resembles, especially in its stature, which can reach 15 m, and by the epithet by which it is sometimes known, "black wattle." *A. decurrens* is more commonly known as "green wattle" and is reputed to possess numerous, inadequately studied, medicinal and pharmacologic properties (www.naturalmedicinefacts.info/plant/acacia-decurrens.html). (a) *A. decurrens*, branch with leaves and flowers, mature seed pod, seed, and trunk; botanical drawing. (By Edward Minchen, in J.H. Maiden and W.S. Campbell, *The Flowering Plants and Ferns of New South Wales*, Part 4 [Sydney: C. Potter, Government Printer, 1896], via www.plantillustrations.org, public domain, www.plantillustrations.org/illustration.php?id_illustration=174974.) (b) *A. decurrens*, blossoming tree. (7/24/2011, Waterfall, NSW Australia: by John Tann, CC BY 2.0, via Flickr, www.flickr.com/photos/31031835@N08/5969318252.)

(a) (b)

FIGURE 2.19 *Eremophila glabra*, tar bush. (a) *E. glabra*, branches with foliage and a red flower. (6/12/2013, Whistler Walk, Gluepot Reserve, South Australia: by Julie Burgher, CC BY-ND 2.0, via Flickr, www.flickr.com/photos/sunphlo/9119897259/.) (b) *E. glabra* subsp. *glabra*, close-up of flowers and leaves. (7/12/2008: by Kevin Thiele, CC BY 2.0, via Flickr, www.flickr.com/photos/66951228@N07/6283410774/.)

FIGURE 2.20 Leaves, buds, and white flowers of the Australian native *Backhousia citriodora*, lemon myrtle. (By KarenHBlack, via Shutterstock, www.shutterstock.com/image-photo/white-flowers-buds-australian-native-lemon-1878056155.)

FIGURE 2.21 *Acacia auriculiformis*, earleaf wattle, is a fast growing, impressive, gnarly acacia, attaining heights of thirty meters, indigenous to Australia, Indonesia, and Papua New Guinea. Extracts have been reputed to possess analgesic properties, along with another acacia, *Acacia ancistrocarpa* (*Australian New Crops Newsletter*, 10, 1998). (a) *A. auriculiformis*, foliage and yellow flowers. (2/9/2017, Muthanallur Amanikere, Karnataka, India: by Ajit Ampalakkad, CC0 1.0, via iNaturalist, www.inaturalist.org/photos/48446273.) (b) *A. auriculiformis*, blossoming tree. (2/9/2017, Muthanallur Amanikere, Karnataka, India: by Ajit Ampalakkad, CC0 1.0, via iNaturalist, www.inaturalist.org/photos/48446265.) (c) *A. auriculiformis*, mature seed pods. (12/27/2018, 18250 N Tamiami Trail, North Fort Myers, FL 33903, USA: by Dr. Alison Northup, CC BY 4.0, via iNaturalist, www.inaturalist.org/photos/30504732.) (d) *A. auriculiformis*, seeds in the seed pod. (12/27/2018, 18250 N Tamiami Trail, North Fort Myers, FL 33903, USA: by Dr. Alison Northup, CC BY 4.0, via iNaturalist, www.inaturalist.org/photos/30504751.)

(a) (b)

FIGURE 2.22 *Acacia kempeana* is called "witchetty bush" because of the large white wood-eating larvae, "witchetty," that enjoy the roots of this multi-stemmed, sprawling shrub. Aboriginal Australians are said to have consumed these larvae as a source of good-quality protein. (a) *A. kempeana* bush with blue-gray leaves. (1/21/2015, Kata-tjuta, Australia: by Kenneth Bader, with permission, via iNaturalist, www.inaturalist.org/photos/1579323.) (b) *A. kempeana* bush in its natural habitat. (1/23/2015, Watarrka National Park, Australia: by Kenneth Bader, with permission, via iNaturalist, www.inaturalist.org/photos/1585788.)

(a) (b)

FIGURE 2.23 *Acacia dealbata*, silver wattle, of southeastern Australia is highly prized as an ornamental throughout the temperate, non-frost regions of the world. Its wood is suitable for fine furniture, while its flowers yield a scent known to perfumery as "mimosa." (a) *A. dealbata* flowers. (3/19/2012: by Miluz, CC BY 2.0, via Flickr, www.flickr.com/photos/35262951@N00/7032636537.) (b) *A. dealbata* leaf and mature seed pods. (6/6/2020, Capoterra, Sardegna, Italia: by Marco Piga, with permission, via iNaturalist, www.inaturalist.org/photos/81851263.)

By the disc diffusion method, extracts of *Acacia dealbata* in hexane, dichloromethane, water, acetone, methanol, and ethanol were tested for their antibiotic effects on *Staphylococcus aureus* and *Escherichia coli*. Ethanolic and acetone extracts delivered the most lethal antibiotic blow to these helpless bacteria (Borges et al. 2020).

Aqueous and methanolic extracts of *Acacia salicina* checked, in order of their sensitivity: *Staphylococcus aureus*, *Salmonella typhimurium*, *Salmonella enteritidis*, *Enterococcus faecalis*, and *Escherichia coli*. Boubaker et al. (2012) and Chatti et al. (2009) found that "total oligomer flavonoids," a partially purified fraction, killed about twice as many bugs as either of the parent fractions.

(a) (b)

FIGURE 2.24 Indigenous Australians know *Acacia salicina*, willow wattle, as *kooba*, extracting fish poison from its bark and smoking its leaves for inebriation. Its wood is very hard, and it was formerly used to make axles for conveyances. (a) *A. salicina* blossoms and honeybees. (2/3/2010: by Sydney Oats, CC BY 2.0, via Flickr, www.flickr.com/photos/57768042@N00/4327195194.) (b) *A. salicina* trees. (8/22/2019, Woodstock-Cleveland-Ross, Reid River, Queensland, Australia: by Elawrey, CC BY 4.0, via iNaturalist, www.inaturalist.org/photos/52725786.)

(a) (b)

FIGURE 2.25 *Senegalia ataxacantha*. The translucent red seed pods or the distinctive flowers, illuminated with backlight, lead to the impression of the bush being alight; hence, this large, multi-stemmed, *irregularly thorned* (= *ataxacantha*) species is commonly called throughout its sub-Saharan African habitats "flame bush." (a) *S. ataxacantha* flowers, buds, and leaves. (2/14/2018, Madwaleni, South Africa: by Peter Warren, CC0, 1.0, via iNaturalist, www.inaturalist.org/photos/14008516.) (b) *S. ataxacantha* branches with leaves and mature seed pods. (5/1/2021, Koedoespoort 456-Jr, Pretoria, 0186, South Africa: by Hildegard Klein, CC BY 4.0, via iNaturalist, www.inaturalist.org/photos/125080570.)

A study of a large selection of Yemeni medicinal plant extracts for possible antibiotic activity using *in vitro* assays included three *Acacia s.l.* species. Among these, the methanolic extract of *Senegalia asak* showed good suppression of the growth of *Staphylococcus aureus* and *Bacillus subtilis*, while *Vachellia nilotica* and *V. tortilis* extracts were effective against *S. aureus*, *B. subtilis*,

FIGURE 2.26 *Senegalia asak*, specimen from Kew's Herbarium. *S. asak* is a small, slender tree or large shrub, attaining heights of 4–10 m. Its gum is used locally in its Ethiopian and other East African habitats. (Collected 4/21/1950, Saudi Arabia, CC BY 3.0, http://specimens.kew.org/herbarium/K000791106 © Board of Trustees of the Royal Botanic Gardens, Kew.)

Pseudomonas aeruginosa, and *Micrococcus flavus* (Al-Fatimi et al. 2007). Amoussa et al. established the antibiotic activities of extracts of *Senegalia ataxacantha* with special attention to a novel chalcone, *acthaside* (2016a), bactericidal at 25 µg/mL with total bacterial annihilation at 50 µg/mL, in a complex triterpenoid matrix (2016b).

In Sudan, where the predominant treatment of infectious maladies is with herbs, 24 different plants recognized as good medicine for infections were selected for treating severe periodontitis, known to often be associated with the anaerobic black-stained pathogen *Porphyromonas gingivalis*. Sixty-two extracts produced with a methanol/ethanol solvent system from the 24 plants were tested *in vitro* against *Porphyromonas gingivalis*. The number two plant part in potency was the bark of *V. tortilis*, just behind *Terminalia laxiflora* with a minimum inhibitory concentration (MIC) of about ½ mg/mL and shortly following the bark of *V. seyal* with a MIC of about 1 mg/mL (Mohieldin et al. 2017).

Ethanolic extracts, especially of the leaves and, to a lesser extent, the flowers, of *Vachellia aroma*, a spunky South American species, showed strong antibiotic activity, possibly bactericidal, against a very wide assortment of both Gram-negative and Gram-positive prokaryotes (Arias et al. 2004). When extracted with ethyl acetate or ethanol, *V. aroma* parts yielded mixtures able to suppress the growth of methicillin-resistant *Staphylococcus aureus* (MRSA), methicillin-sensitive *S. aureus* (MSSA) and methicillin-resistant *S. epidermidis*, however, the aqueous extracts of the same tree parts were essentially inactive against these *Staphylococcus* strains (Mattana et al. 2010).

"Quorum sensing" is the ability of an organism (most often used in reference to bacteria) to sense the density of its population and to regulate gene expression accordingly. It is an important means for optimizing the growth of its population, so interventions able to disrupt this bacterial capability constitute a potential mechanism for antibiotic control. In this context, crude extracts of *Senegalia nigrescens*, along with some of the pure compounds isolated from these extracts, namely a novel ent-kaurene diterpenoid (ent-kaur-15-en-18,20-diol), for the first time found in a plant species, ent-kaur-15-en-18-ol, 30-hydroxylup-20(29)-en-3β-ol, 3β-hydroxy-20(29)-en-lupan-30-al, lupeol, stigmasterol, a long-chain alcohol (tetracosan-1-ol), and three flavonoids (melanoxetin, quercetin, and quercetin-3-O-methyl ether) inhibited the quorum sensing capability of *Chromobacterium violaceae* (Bodede et al. 2018).

(a) (b)

FIGURE 2.27 *Vachellia tortilis*, the "umbrella thorn acacia," is indigenous to Israel and the Middle East and a possible candidate for the acacia of the Bible, the wood for the Ark of the Covenant, but it grows in most parts of Africa as well. (a) *V. tortilis* tree in the sandy Arava Valley. (3/29/2019, Sheizaf, Arava Valley, Israel: by Kristof Zyskowski, CC BY 4.0, via iNaturalist, www.inaturalist.org/photos/36343949.) (b) *V. tortilis*, curved seed pods. (3/29/2019, Sheizaf, Arava Valley, Israel: by Kristof Zyskowski, CC BY 4.0, via iNaturalist, www.inaturalist.org/photos/36339566.)

(a) (b) (c)

FIGURE 2.28 Another chief contender for the biblical *shita* wood used to fashion the Ark of the Covenant, as well as coffins for Egyptian royalty, *Vachellia seyal* (red acacia) shares much of the same habitat as *V. tortilis*, yet it often grows in damp valleys. It has multiple folk medicinal uses associated with its antibiotic potency for treatments ranging from dysentery to leprosy. (a) *V. seyal* tree. (3/23/2013: by TreeWorld Wholesale, CC BY 2.0, via Flickr, www.flickr.com/photos/treeworld/34721375172/.) (b) *V. seyal* branches. (7/29/2016: by TreeWorld Wholesale, CC BY 2.0, via Flickr, www.flickr.com/photos/treeworld/34844850406/.) (c) *Acacia seyal* Delile (= *V. seyal*) branches with flowers, leaves, thorns, seeds, and seed pods; botanical drawing. (In F. G. Hayne, *Getreue Darstellung und Beschreibung der in der Arzneykunde gebräuchlichen Gewächse* [Berlin, 1805–46], vol. 10, t. 30, public domain, via www.plantillustrations.org, www.plantillustrations.org/illustration.php?id_illustration=353413.)

(a) (b)

FIGURE 2.29 *Vachellia aroma. Aromita* and *espenillo* are common, revealingly descriptive names for this small, spiky tree of Peru, Chile, Argentina, and Paraguay. It is known as a source for making honey, for crafting wooden implements, and as a respectable fuel for burning. (a) *V. aroma* branches with buds, flowers, and leaves. (10/8/2017, Banda, Santiago del Estero, Argentina: by Jose Luis Navarro, CC BY 4.0, via iNaturalist, www.inaturalist.org/photos/146918560.) (b) *V. aroma* seed pods. (2/23/2019, Boquerón, Paraguay: by Étienne Lacroix-Carignan, CC0 1.0, via iNaturalist, www.inaturalist.org/photos/32360370.)

(a) (b) (c)

FIGURE 2.30 *Senegalia nigrescens*, the knobthorn tree, is tall, as high as 18 m, with foliage out of reach for most mammals. This is a favorite of African giraffes, possibly the only mammals able to utilize it as a nutritional source. (a) *S. nigrescens* trees, habitat. (1/25/2020, Kruger Park, South Africa: by Jens-Christian Svenning, CC0 1.0, via iNaturalist, www.inaturalist.org/photos/60416364.) (b) *S. nigrescens* trunk with characteristic "knobs." (9/18/2016, Ehlanzeni, South Africa: by Lallen, CC0 1.0, via iNaturalist, www.inaturalist.org/photos/66578373.) (c) *S. nigrescens* branch with leaves and white flowers. (10/6/2012, Pretoria, South Africa: by VanLap Hoàng, CC BY 2.0, via Flickr, www.flickr.com/photos/vanlaphoang1945/9311486881/.)

Campylobacter species, specifically *C. jejuni* and *C. coli*, may cause serious pathology, ranging from severe diarrhea to Guillain-Barre syndrome, a neurological disorder. An extract of *Vachellia farnesiana* killed these bacteria at a minimal bactericidal concentration (MBC) of 0.3 mg/mL (Castillo et al. 2011). Methanolic, ethanolic, and aqueous extracts of *V. farnesiana* parts were tested against *Vibrio cholera* and shown to disrupt the prokaryote's cell membrane, resulting in mortal damage, via enhanced membrane permeability, lowering of cytoplasmic pH, cell membrane hyperpolarization, and decreased intracellular ATP. Methanolic extracts were the most potent (Sánchez et al. 2010).

The non-polar elements of *V. farnesiana* chloroform and hexane extracts were isolated and identified as tetracosanoic acid (2S)-2,3-dihydroxypropyl ester and (3β,22E)-estigmasta-5,22-dien-3-yl β-D-glucopyranoside, though the polar methanolic extract containing methyl gallate, gallic acid, (3β,22E)-estigmasta-5,22-dien-3-yl β-D-glucopyranoside, (2S) naringenin 7-O-β-glucopyranoside (prunin), pinitol, and sucrose was significantly more active against *Mycobacterium tuberculosis* and dysentery-associated *Campylobacter jejuni*, *Shigella flexneri*, *Salmonella enteritidis*, *Yersinia enterocolitica*, and the enterohemorrhagic *E. coli* (Hernández-García et al. 2019).

In a screening of solvent extracts of ten different South African trees for antimicrobial outcome, ethyl acetate of the roots of *Vachellia sieberiana* was most efficacious, with an MIC of 92 μg/mL against *Staphylococcus aureus* and *E. coli* (Eldeen et al. 2005).

An *in situ* clinical study (Jayashankar et al. 2011) found that *Senegalia chundra*, either by itself or with eight other herbs in a dentifrice, decreased plaque formation and the growth of both aerobic and anaerobic oral cavity bacteria by a large amount.

(a) (b) (c)

FIGURE 2.31 *Vachellia farnesiana*, also known as sweet acacia, mimosa bush, or needle bush, is a small, usually deciduous but sometimes evergreen tree or large bush, reaching a height of 9 m. The plant is native to the Dominican Republic, where it was first discovered, and grows throughout Central America, Mexico, Hawaii, and Australia. Its blossoms yield a perfume named *cassie*, while the seeds are traditionally used to poison rabid dogs and the leaves for killing protozoan parasites. Farnesol, a sterol precursor, was discovered and named after this plant. (a) The *V. farnesiana* tree bears both flowers and seed pods at the same time. (3/2/2020, Calle Río la Silla, Guadalupe, Nuevo León, Mexico: by Adriana Nelly Correa Sandoval, CC BY 4.0, via iNaturalist, www.inaturalist.org/photos/13848708.) (b) *V. farnesiana* branches with yellow flowers, flower buds, leaves, and mature and immature seed pods. (10/6/2008, Decatur Ave Sand Island, Midway Atoll, Hawaii: by Forest & Kim Starr, CC BY 4.0, via Starr Environmental, www.starrenvironmental.com/images/image/?q=24893540446.) (c) *V. farnesiana*, open mature pods and seeds. (10/20/2005, Honokanaia, Kahoolawe, Hawaii: by Forest & Kim Starr, CC BY 4.0, via Starr Environmental, www.starrenvironmental.com/images/image/?q=24553247150.)

(a)

(b)

(c)

FIGURE 2.32 *Vachellia sieberiana*, paperbark thorn acacia, may reach heights of 25 m and a width of up to 1.8 m. Aqueous decoctions of the bark are employed in traditional African medicine for childhood fevers, as an astringent, and as treatments for social diseases including gonorrhea. (a) *V. sieberiana* tree. (5/1/2021, Koedoespoort 456-Jr, Pretoria, 0186, South Africa: by Hildegard Klein, CC BY 4.0, via iNaturalist, www.inaturalist.org/photos/125078091.) (b) *V. sieberiana* gets its English name "paperbark acacia" from its flaky bark. (8/7/2021, uMngeni Local Municipality, South Africa: by Benjamin Fredlund, CC BY 4.0, via iNaturalist, www.inaturalist.org/photos/149717259.) (c) *V. sieberiana*, foliage and seed pods. (6/11/2020, Bishop Bird Park, Centurion, Gauteng, South Africa: by Matthew Fainman, CC BY 4.0, via iNaturalist, www.inaturalist.org/photos/78114510.)

(a) (b)

FIGURE 2.33 The wood of *Senegalia chundra*, red cutch, is prized in shipbuilding for its water-resistant properties and strength, while the tree, which can grow to 15 m, is a favorite nesting milieu for the scaly-breasted munia, *Lonchura punctulata*, especially in Asia, India, and the Indian Ocean region. (a) *S. chundra* branches with buds, flowers, and leaves. (5/18/2017, Bangalore Urban, Karnataka, India: by Ajit Ampalakkad, CC0 1.0, via iNaturalist, www.inaturalist.org/photos/157726750.) (b) *S. chundra* branch with thorns. (5/18/2017, Bangalore Urban, Karnataka, India: by Ajit Ampalakkad, CC0 1.0, via iNaturalist, www.inaturalist.org/photos/157726672.)

FIGURE 2.34 *Lonchura punctulata*, scaly-breasted munia, or spotted munia. (1/10/2019, Durgapur, Mirsharai Upazila, Chittagong, Bangladesh: by Tareq Uddin Ahmed, CC BY 2.0, via Flickr, www.flickr.com/photos/90769516@N06/49416567726/.)

The effects of adding *Acaciella angustissima* leaves to the diet of ruminant livestock on the suppression of certain microorganisms were studied. *Butyrivibrio fibrisolvens*, a common rumen resident, was particularly sensitive to dietary *A. angustissima* leaves, while *Selenomonas* spp. were resistant. Differences in bacterial sensitivity may have been due to general differences in sensitivity to tannins, though the possible influence of other compounds in the leaves on the antibiotic effect was not excluded (Krause et al. 2004).

(a)

(b)

(c)

FIGURE 2.35 Extracts of the roots or bark of *Acaciella angustissima*, also known as prairie acacia or white ball acacia, are a frequent additive to pulque and other alcoholic beverages in Mexico, the Caribbean, and South America, where this bush is indigenous. It likes moisture, and so its adoption in arid zones is limited. It is used ethnomedically for disorders of the digestion and other settings where an infectious component occurs. (a) *A. angustissima* branch with flowers and leaves. (8/25/2017, Santa Cruz County, Arizona, USA: by Don Loarie, CC BY 4.0, via iNaturalist, www.inaturalist.org/photos/10088193.) (b) *A. angustissima*, bark. (3/26/2019, Cadereyta de Montes, Qro., México: by José Belem Hernández Díaz, CC BY 4.0, via iNaturalist, www.inaturalist.org/photos/33624978.) (c) *Acacia angustissima* (Mill.) Kuntze (= *Acaciella angustissima*), botanical drawing. (In *Annals of the Missouri Botanical Garden* 37 (1950): p. 269, t. 97, via www.plantillustrations.org, public domain, www.plantillustrations.org/illustration.php?id_illustration=116739.)

Senegalia mellifera bark was extracted with methanol or a 1:1 methanol/dichloromethane mixture; both extracts were potently antimicrobial against *Staphylococcus aureus* and *Pseudomonas aeruginosa* (Mutai et al. 2009).

In a study of the hot water and methanolic extracts from 26 medicinal plants of the Yemeni island of Socotra, one Socotra-endemic acacia, *Acacia pennivenia*, was among the best in *in vitro* antimicrobial activity against both Gram-positive and Gram-negative bacteria and several *Streptococcus* variants (Mothana et al. 2009).

FIGURE 2.36 *Senegalia mellifera*, blackthorn. (a) *S. mellifera* ssp. *detinens*, bush with white flowers. (9/5/2016, Ghanzi, Botswana: by Lallen, CC0 1.0, via iNaturalist, www.inaturalist.org/photos/51073963.) (b) *S. mellifera* ssp. *detinens*, bark, flower buds, and the characteristic black thorns. (8/25/2018, Ghanzi, Botswana: by Robert Taylor, CC BY 4.0, via iNaturalist, www.inaturalist.org/photos/53186650.) (c) *Acacia mellifera* (= *S. mellifera*) seed pods. (12/2/2018, Mokala NP, Northern Cape, South Africa: by Bernard Dupont, with permission, via Flickr, www.flickr.com/photos/berniedup/33268114378.)

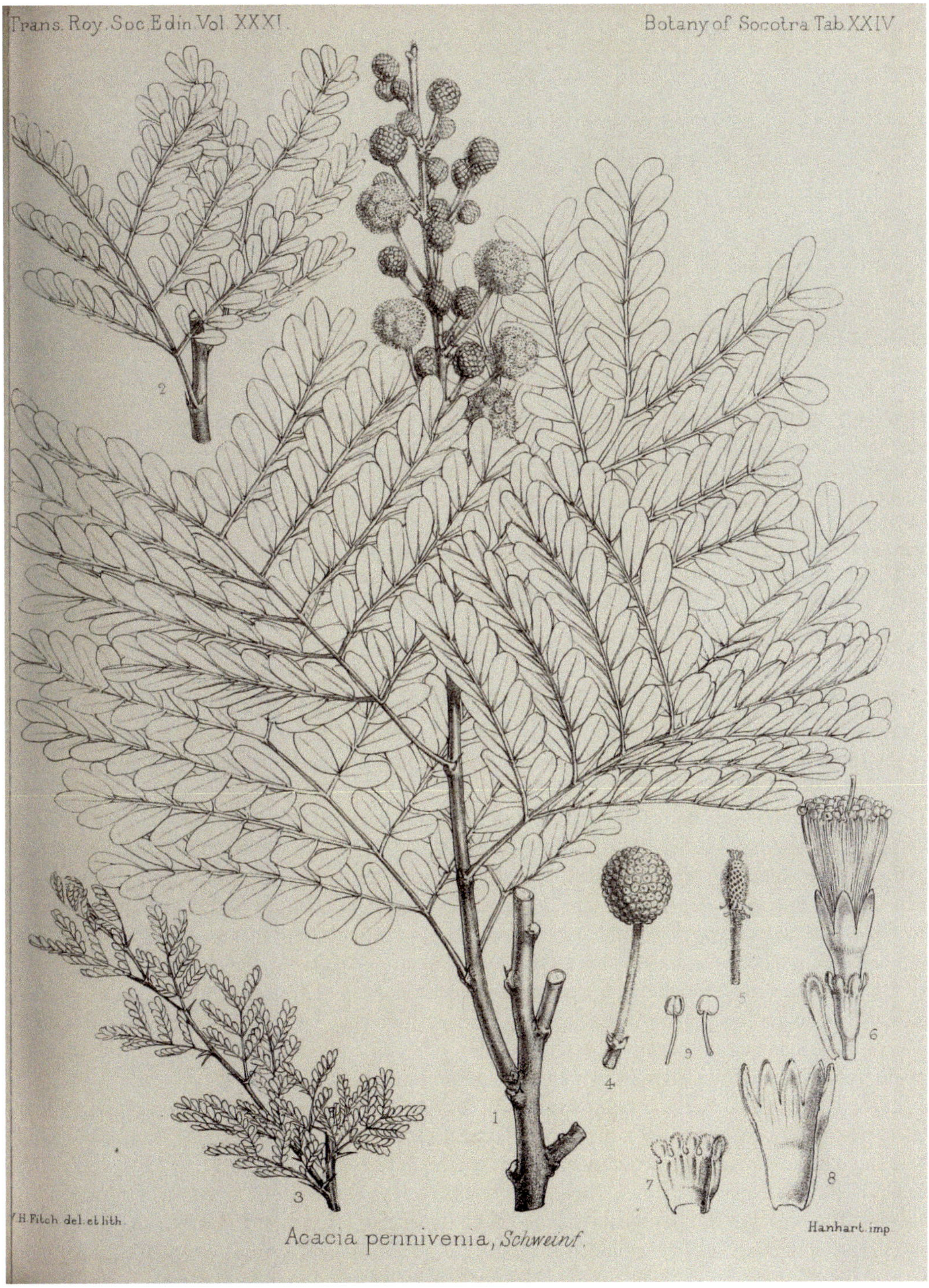

FIGURE 2.37 *Acacia pennivenia* is endemic to Socotra, the Yemeni island of genetic uniqueness. It is rare, difficult to access, and endangered. *A. pennivenia*, botanical drawing. (By W. H. Fitch, in I. B. Balfour, *Botany of Socotra*, *Transactions of the Royal Society of Edinburgh* 32 [Edinburgh: R. Grant, 1888], table 24, public domain, via www.plantillustrations.org/illustration.php?id_illustration=322345.)

(a)

(b)

FIGURE 2.38 *Acacia aneura*, mulga tree, is native to the "mulga regions" of southwest Australia, which took their name from this tree that inhabits those distinctive shrublands. The seeds are edible and can be used to make seed cakes. A gall that may grow on a tree, a "mulga apple," was eaten by Aboriginal peoples. The gum of this thornless acacia is sweet like honey, a mammalian treat. (a) *A. aneura* tree in its habitat in Hawaii. (4/2/2007, Kaumalapau Hwy, Lanai, Hawaii: by Forest & Kim Starr, CC BY 4.0, via Starr Environmental, www.starrenvironmental.com/images/image/?q=24590755120.) (b) *A. aneura* branch with buds, flowers, and leaves. (4/2/2007, Kaumalapau Hwy, Lanai, Hawaii: by Forest & Kim Starr, CC BY 4.0, via Starr Environmental, www.starrenvironmental.com/images/image/?q=24792809021.)

When the aerial parts of seven so-called "multipurpose" African trees were extracted and tested for possible antibacterial and antiprotozoal activity aimed at veterinary applications, the one acacia in the series, *Acacia aneura*, provided the highest toxicity to bacteria inhabiting the rumen (Newbold et al. 1997).

Senegalia burkei was one of 23 native South African plant species utilized by Zulus for the treatment of diarrhea that was tested for its antibiotic activity. Unlike most of the species, where aqueous extracts were weak or without effect, the water extract of *S. burkei* retained its efficacy at a low MIC against *Bacillus cereus*, *Enterococcus faecalis*, *Escherichia coli*, *Proteus vulgaris*, *Salmonella typhimurium*, *Shigella flexneri*, and *Staphylococcus aureus* (van Vuuren et al. 2015). *S. burkei* is an officially recognized anti-diarrheal drug plant (de Wet et al. 2010).

Owayss et al. (2019) found that honey made from the flowers of *Acacia gerrardii* was much more effective as an antibiotic against pathogenic Gram-positive bacteria *Bacillus cereus* and *Staphylococcus aureus* and Gram-negative bacteria *Escherichia coli* and *Salmonella enteritidis* than honey made from *Ziziphus spina-christi*, which is a common tree in Saudi Arabia.

In a study of polar and non-polar extracts of 22 African medicinal plants, the highest antibacterial potency was achieved by a methanolic extract of the aerial parts of *Vachellia karroo*, which was more effective against Gram-positive (*Staphylococcus aureus*, *Enterococcus faecalis*) than Gram-negative (*Escherichia coli*, *Pseudomonas aeruginosa*, *Klebsiella pneumoniae*) pathogenic species. The presence of tannins and flavonoids in the extracts was postulated as a causative factor in the antibiotic responses (Madureira et al. 2012). An ethnomedical review found it to be extensively utilized in its native milieu for colds, diarrhea, dysentery, flu, malaria, sexually transmitted infections (STIs), wounds, colic, and ethnoveterinary medicine and to possess in its leaves and roots flavonoids, phenols, phytosterols, proanthocyanidin, tannin, terpenes, and minerals (Maroyi 2017). In an *in vitro* study, ethanolic and dichloromethane extracts of *V. karroo* bark effected a 1 mg/mL MIC against two Gram-positive (*Bacillus subtilis* and *Staphylococcus aureus*) and three Gram-negative (*Neisseria gonorrhoeae*, *Escherichia coli*, and *Klebsiella pneumoniae*) bacteria (Mulaudzi et al. 2011).

(a) (b)

FIGURE 2.39 *Senegalia burkei*, black monkey thorn, is a true South African native. Large and sprawling, its white flowers give way to bright red seed pods (i.e., fruits). Other names include *swartapiesdoring* (Afrikaans); *umkhaya wehlalahlathi*, *umbabampala* (isiZulu); *umkhaya* (siSwati); *mokgwa*, *mokoba* (Setswana); *mokwaripa*, *mongangatau* (Sesotho sa Leboa); and *munanga* (Tshivenda) (pza.sanbi.org/senegalia-burkei). *S. burkei* has also been successfully cultivated as a bonsai, or miniature tree. (a) *S. burkei* flowers and leaves. (11/26/2013, H4–1 Road West of Lower Sabie, Kruger National Park, South Africa: by Bernard Dupont, with permission, via Flickr, www.flickr.com/photos/berniedup/50689451468/.) (b) *S. burkei* tree, trunk and branches. (1/7/2006, Haka Game Park near Harare, Zimbabwe: by Markus de Klerk, CC BY 4.0, via iNaturalist, www.inaturalist.org/photos/151112.)

(a) (b)

FIGURE 2.40 *Vachellia gerrardii*, red acacia. This inhabitant of Israel and the neighboring Middle Eastern deserts possesses serious thorns. Its wood is valuable for small projects. The inner bark is chewed or boiled in aqueous decoctions as a remedy for childhood diarrhea, cough, and sore throat, and possesses emetic properties as well. (a) *V. gerrardii* tree. (2/16/2019, Living Desert Zoo and Gardens, Acacia Collection, Palm Desert, Riverside County, CA: by K M, CC BY 2.0, via Flickr, www.flickr.com/photos/131880272@N06/40207479013/.) (b) *V. gerrardii*, trunk with the red bark from which the tree has got its English name. (2/16/2019, Living Desert Zoo and Gardens, Acacia Collection, Palm Desert, Riverside County, CA: by K M, CC BY 2.0, via Flickr, www.flickr.com/photos/131880272@N06/46258244995.)

(c) (d) (e)

FIGURE 2.40 (Continued) (c) *Acacia gerrardii* Benth. (= *V. gerrardii* [Benth.] P.J.H.Hurter), branches with white flowers and foliage. (2/15/2008: by Paul Latham, with permission, via Flickr, www.flickr.com/photos/36517976@N06/3446212727/.) (d) *V. gerrardii* var. *gerrardii*, branch with foliage and thorns. (1/10/2015, near Hamasha Camp, Leshiba Wilderness, South Africa: by Wynand Uys, CC BY 4.0, via iNaturalist, www.inaturalist.org/photos/15790885.) (e) *V. gerrardii* flower. The honey of the flower has noted antibiotic qualities. (1/9/2013, Walter Sisulu National Botanic Garden, Roodepoort, Johannesburg, South Africa: by Charles Stirton, with permission, via iNaturalist, www.inaturalist.org/photos/15326253.)

(a) (b)

FIGURE 2.41 *Ziziphus spina-christi*, Christ's thorn jujube. (a) *Z. spina-christi* flowers. (12/15/2018, Sharjah, United Arab Emirates: by Jacky Judas, CC BY 4.0, via iNaturalist, www.inaturalist.org/photos/46616085.) (b) *Z. spina-christi*, fruit. (3/28/2019, Ein Gedi, Judean Desert, Israel: by Kristof Zyskowski, CC BY 4.0, via iNaturalist, www.inaturalist.org/photos/36182171.)

(a) (b) (c)

FIGURE 2.42 *Vachellia karroo*, also known as "sweet thorn," "cockspur thorn," or "karoo thorn," is a native southern African bush or small tree attaining maximum heights of 12 m and preferring arid regions. The edible gum from the wood is a treat for the bush-baby (*Galago* sp.), and is also another name for the tree (sweet gum) and a natural material for the production of candies. The different plant parts are used in traditional medicine for their emollient and astringent properties. (a) *V. karroo* tree with yellow flowers. (11/26/2013, H4–1 Road East of Skukuza, Kruger Natural Park, South Africa: by Bernard Dupont, with permission, via Flickr, www.flickr.com/photos/berniedup/50692158438.) (b) *V. karroo* branch with flowers, buds, leaves, and thorns. (11/24/2020, Roodeplaat 5, City of Tshwane Metropolitan Municipality, South Africa: by Hildegard Klein, CC BY 4.0, via iNaturalist, www.inaturalist.org/photos/105572523.) (c) *V. karroo* bark, with sweet-tasting gum seeping from cracks in it. (7/4/2020, Weenen Game Reserve, Uthukela DC, South Africa: by Alan Manson, CC BY 4.0, via iNaturalist, www.inaturalist.org/photos/107920448.)

Twenty-eight different plants used by the Nahua Indians in Mexico for medicinal purposes were tested for their possible antibiotic effect against pathogenic *Staphylococcus aureus* and *Yersinia enterocolitica*, and nonpathogenic *Escherichia coli* and *Micrococcus luteus*. Bork et al. (1996) found that out of the 28 plants, *Vachellia cornigera, Cuscuta tinctoria, Ludwigia octovalvis, Lysiloma divaricata,* and *Tithonia diversifolia* were the five most effective against the microorganisms.

(a) (b) (c)

FIGURE 2.43 *Vachellia cornigera*, bullhorn acacia, is so named, both in English (bullhorn) and Latin (*cornigera*), on account of its swollen and hollow horn-like thorns, which provide ants with lodging, while the tree's sweet nectar makes for board. The ants help the tree fend off other insects and herbivorous animals. Over time, the inducible protective secretion of extrafloral nectar to feed ants on an as-needed basis, became constitutive, i.e., permanent (Heil et al. 2004). In Mexico, where the plant grows, the beautiful thorns may be fashioned into necklaces for dolls, and they have been used for therapeutic needling in traditional Mayan acupuncture. (a) *V. cornigera* branch with leaves and thorns. (4/3/2015, Cayo, Belize: by Feroze Omardeen, CC BY 2.0, via Flickr, www.flickr.com/photos/sucriertt/17171585962.) (b) *V. cornigera* branch with leaves and seed pods. (3/3/2018, Costa Rica: by Andy Blackledge, CC BY 2.0, via Flickr, www.flickr.com/photos/hockeyholic/27185099998.) (c) *V. cornigera*, botanical drawing. (By M. Smith and J. N. Fitch in J. D. Hooker, *Curtis's Botanical Magazine* 121 [= ser. 3, v. 51] [London: L. Reeve & Co., 1895], Tab. 7395, public domain, via Flickr [6/21/2017: by Biodiversity Heritage Library, public domain, www.flickr.com/photos/biodivlibrary/34596721734/].)

FIGURE 2.44 *Galago moholi* (Nocturnal Lesser Bushbaby) sitting in a tree. This small and smart non-simian primate loves the sweet gum of *V. karroo*. (South Africa: by EcoPrint, via Shutterstock, www.shutterstock.com/image-photo/nocturnal-lesser-bushbaby-galago-moholi-sitting-217008154.)

Not to be confused with another famous resident of the Australian gum tree, the original gum drops were a collection of bark exudates, the Australian equivalent of "gum arabic," and the famous bird loved them as much as the bushbabies love their "gum drops."

> . . . Kookaburra sits in the old gum tree,
> Eating all the gumdrops he can see . . .
> Laugh Kookaburra, laugh Kookaburra, gay your life must be.
>
> —"Kookaburra" by Marion Sinclair (1932)

FIGURE 2.45 *Dacelo novaeguineae*, laughing kookaburra. The laughing kookaburra ... love that gum tree! (2/27/2012: by Brisbane City Council, CC BY 2.0, via Flickr, www.flickr.com/photos/brisbanecitycouncil/8025631404.)

REFERENCES

Aadil, K.R., S.I. Mussatto, and H. Jha. 2018. Synthesis and characterization of silver nanoparticles loaded poly(vinyl alcohol)-lignin electrospun nanofibers and their antimicrobial activity. *Int J Biol Macromol* 120(Pt A): 763–7.

Akter, K., E.C. Barnes, J.J. Brophy, et al. 2016. Phytochemical profile and antibacterial and antioxidant activities of medicinal plants used by Aboriginal people of New South Wales, Australia. *Evid Based Complement Alternat Med* 2016: 4683059.

Al-Fatimi, M., M. Wurster, G. Schröder, and U. Lindequist. 2007. Antioxidant, antimicrobial and cytotoxic activities of selected medicinal plants from Yemen. *J Ethnopharmacol* 111(3): 657–66.

Almas, K. 2001. The antimicrobial effects of seven different types of Asian chewing sticks. *Odontostomatol Trop* 24(96): 17–20.

Amin, M., F. Anwar, F. Naz, T. Mehmood, and N. Saari. 2013. Anti-helicobacter pylori and urease inhibition activities of some traditional medicinal plants. *Molecules* 18(2): 2135–49.

Amoussa, A.M., M. Bourjot, L. Lagnika, C. Vonthron-Sénécheau, and A. Sanni. 2016a. Acthaside: A new chromone derivative from *Acacia ataxacantha* and its biological activities. *BMC Complement Altern Med* 16(1): 506.

Amoussa, A.M., L. Lagnika, M. Bourjot, C. Vonthron-Senecheau, and A. Sanni. 2016b. Triterpenoids from *Acacia ataxacantha* DC: Antimicrobial and antioxidant activities. *BMC Complement Altern Med* 16(1): 284.

Ansari, M.A., H.M. Khan, A.A. Khan, S.S. Cameotra, Q. Saquib, and J. Musarrat. 2014. Gum arabic capped-silver nanoparticles inhibit biofilm formation by multi-drug resistant strains of *Pseudomonas aeruginosa*. *J Basic Microbiol* 54(7): 688–99.

Arias, M.E., J.D. Gomez, N.M. Cudmani, M.A. Vattuone, and M.I. Isla. 2004. Antibacterial activity of ethanolic and aqueous extracts of *Acacia aroma* Gill. ex Hook et Arn. *Life Sci* 75(2): 191–202.

Bai, S., P. Bharti, L. Seasotiya, A. Malik, and S. Dalal. 2015. *In vitro* screening and evaluation of some Indian medicinal plants for their potential to inhibit Jack bean and bacterial ureases causing urinary infections. *Pharm Biol* 53(3): 326–33.

Baien, S.H., J. Seele, T. Henneck, et al. 2020. Antimicrobial and immunomodulatory effect of gum arabic on human and bovine granulocytes against *Staphylococcus aureus* and *Escherichia coli*. *Front Immunol* 10: 3119.

Bajpai, S.K., M. Jadaun, and S. Tiwari. 2016. Synthesis, characterization and antimicrobial applications of zinc oxide nanoparticles loaded gum acacia/poly(SA) hydrogels. *Carbohydr Polym* 153: 60–5.

Bajpai, S.K., and M. Kumari. 2015. A green approach to prepare silver nanoparticles loaded gum acacia/poly(acrylate) hydrogels. *Int J Biol Macromol* 80: 177–88.

Barad, M.K., K.B. Ishnava, and J.B. Chauhan. 2014. Anticariogenic activity and phytochemical studies of crude extract from some Indian plant leaves. *J Intercult Ethnopharmacol* 3(2): 85–90.

Bartzoka, E.D., H. Lange, G. Poce, and C. Crestini. 2018. Stimuli-responsive tannin-Fe^{III} hybrid microcapsules demonstrated by the active release of an anti-tuberculosis agent. *ChemSusChem* 11(22): 3975–91.

Bhatnagar, M., L. Parwani, V. Sharma, J. Ganguli, and A. Bhatnagar. 2013. Hemostatic, antibacterial biopolymers from *Acacia arabica* (Lam.) Willd. and *Moringa oleifera* (Lam.) as potential wound dressing materials. *Indian J Exp Biol* 51(10): 804–10.

Bisi-Johnson, M.A., C.L. Obi, B.B. Samuel, J.N. Eloff, and A.I. Okoh. 2017. Antibacterial activity of crude extracts of some South African medicinal plants against multidrug-resistant etiological agents of diarrhoea. *BMC Complement Altern Med* 17(1): 321.

Bodede, O., S. Shaik, H. Chenia, P. Singh, and R. Moodley. 2018. Quorum sensing inhibitory potential and in silico molecular docking of flavonoids and novel terpenoids from *Senegalia nigrescens*. *J Ethnopharmacol* 216: 134–46.

Borges, A., H. José, V. Homem, and M. Simões. 2020. Comparison of techniques and solvents on the antimicrobial and antioxidant potential of extracts from *Acacia dealbata* and *Olea europaea*. *Antibiotics (Basel)* 9(2): 48.

Bork, P.M., M.L. Schmitz, C. Weimann, M. Kist, and M. Heinrich. 1996. Nahua Indian medicinal plants (Mexico): Inhibitory activity on NF-κB as an anti-inflammatory model and antibacterial effects. *Phytomedicine* 3(3): 263–9.

Boubaker, J., H. Ben Mansour, K. Ghedira, and L. Chekir Ghedira. 2012. Polar extracts from (Tunisian) *Acacia salicina* Lindl. Study of the antimicrobial and antigenotoxic activities. *BMC Complement Altern Med* 12: 37.

Castillo, S.L., N. Heredia, J.F. Contreras, and S. García. 2011. Extracts of edible and medicinal plants in inhibition of growth, adherence, and cytotoxin production of *Campylobacter jejuni* and *Campylobacter coli*. *J Food Sci* 76(6): M421–M426.

Chandra Shekar, B.R., R. Nagarajappa, R. Singh, S. Suma, and R. Thakur. 2016. Antimicrobial efficacy of the combinations of *Acacia nilotica, Murraya koenigii* (Linn.) Sprengel, *Eucalyptus*, and *Psidiumguajava* on primary plaque colonizers: An *in vitro* study. *Indian J Dent Res* 27(4): 415–20.

Chatti, I.B., I. Limem, J. Boubaker, et al. 2009. Phytochemical, antibacterial, antiproliferative, and antioxidant potentials and DNA damage-protecting activity of *Acacia salicina* extracts. *J Med Food* 12(3): 675–83.

Chew, Y.L., E.W. Chan, P.L. Tan, Y.Y. Lim, J. Stanslas, and J.K. Goh. 2011. Assessment of phytochemical content, polyphenolic composition, antioxidant and antibacterial activities of Leguminosae medicinal plants in Peninsular Malaysia. *BMC Complement Altern Med* 11: 12.

Chong, W.S.C., G.A. Dykes, and R. Coorey. 2019. Chemical composition and anti-nutritional profiling of wattle (*Acacia cyclops, Acacia microbotrya* and *Acacia victoriae*) seed originating from Western Australia. *Plant Foods Hum Nutr* 74(4): 518–24.

Chopra, M., M. Bernela, P. Kaur, A. Manuja, B. Kumar, and R. Thakur. 2015. Alginate/gum acacia bipolymeric nanohydrogels—promising carrier for zinc oxide nanoparticles. *Int J Biol Macromol* 72: 827–33.

Clark, D.T., M.I. Gazi, S.W. Cox, B.M. Eley, and G.F. Tinsley. 1993. The effects of *Acacia arabica* gum on the in vitro growth and protease activities of periodontopathic bacteria. *J Clin Periodontol* 20(4): 238–43.

Clement, B., C. Goff, and D. Forbes. 2007. Toxic amines and alkaloids from *Acacia rigidula*. *Phytochemistry* 49: 1377–80.

Dabur, R., A. Gupta, T.K. Mandal, et al. 2007. Antimicrobial activity of some Indian medicinal plants. *Afr J Tradit Complement Altern Med* 4(3): 313–18.

Daniels, R.P. 1949. Penicillin suspended in acacia powder. *Hosp Corps Q* 22(1): 11.

Dashtdar, M., M.R. Dashtdar, B. Dashtdar, G.A. Khan, and K. Kardi. 2016. Phenol-rich compounds sweet gel: A statistically more effective antibiotic than cloxacillin against *Pseudomonas aeruginosa*. *J Pharmacopuncture* 19(3): 246–52.

de Wet, H., M.N. Nkwanyana, and S.F. van Vuuren. 2010. Medicinal plants used for the treatment of diarrhoea in northern Maputaland, KwaZulu-Natal Province, South Africa. *J Ethnopharmacol* 130(2): 284–9.
Eldeen, I.M., E.E. Elgorashi, and J. van Staden. 2005. Antibacterial, anti-inflammatory, anti-cholinesterase and mutagenic effects of extracts obtained from some trees used in South African traditional medicine. *J Ethnopharmacol* 102(3): 457–64.
Eldeen, I.M., F.R. Van Heerden, and J. Van Staden. 2010. *In vitro* biological activities of niloticane, a new bioactive cassane diterpene from the bark of *Acacia nilotica* subsp. *kraussiana*. *J Ethnopharmacol* 128(3): 555–60.
Eltayeb, I.B., A.I. Awad, M.A. Elderbi, and S.A. Shadad. 2004. Effect of gum arabic on the absorption of a single oral dose of amoxicillin in healthy Sudanese volunteers. *J Antimicrob Chemother* 54(2): 577–8.
Escárcega-González, C.E., J.A. Garza-Cervantes, A. Vázquez-Rodríguez, et al. 2018. In vivo antimicrobial activity of silver nanoparticles produced via a green chemistry synthesis using *Acacia rigidula* as a reducing and capping agent. *Int J Nanomedicine* 13: 2349–23.
Farag, M.A., D.A. Al-Mahdy, R. Salah El Dine, et al. 2015. Structure-activity relationships of antimicrobial gallic acid derivatives from pomegranate and acacia fruit extracts against potato bacterial wilt pathogen. *Chem Biodivers* 12(6): 955–62.
Gupta, D., and R.K. Gupta. 2015. Investigation of antibacterial efficacy of *Acacia nilotica* against salivary mutans streptococci: A randomized control trial. *Gen Dent* 63(1): 23–7.
Heil, M., S. Greiner, H. Meimberg, et al. 2004. Evolutionary change from induced to constitutive expression of an indirect plant resistance. *Nature* 430(6996): 205–8.
Hendra, R., A. Willis, and P.A. Keller. 2019. Phytochemical studies on the Australian native plant species *Acacia pycnantha* and *Jacaranda mimosifolia* D.Don. *Nat Prod Res* 33(14): 1997–2003.
Hernández-García, E., A. García, E. Garza-González, et al. 2019. Chemical composition of *Acacia farnesiana* (L) wild fruits and its activity against *Mycobacterium tuberculosis* and dysentery bacteria. *J Ethnopharmacol* 230: 74–80.
Hitchcock, M. 2012. *A Celebration of Wattle: Australia's National Emblem*. Dural Delivery Centre: Rosenberg Publishing.
Jæger, D., B.S. Simpson, C.P. Ndi, et al. 2018. Biological activity and LC-MS/MS profiling of extracts from the Australian medicinal plant *Acacia ligulata* (Fabaceae). *Nat Prod Res* 32(5): 576–81.
Jamila, N., N. Khan, I.M. Hwang, et al. 2020. Characterization of natural gums via elemental and chemometric analyses, synthesis of silver nanoparticles, and biological and catalytic applications. *Int J Biol Macromol* 147: 853–66.
Jayashankar, S., G.J. Panagoda, E.A. Amaratunga, K. Perera, and P.S. Rajapakse. 2011. A randomised double-blind placebo-controlled study on the effects of a herbal toothpaste on gingival bleeding, oral hygiene and microbial variables. *Ceylon Med J* 56(1): 5–9.
Jelassi, A., M. Hassine, M. Besbes Hlila, and H. Ben Jannet. 2017. Chemical composition, antioxidant properties, α-glucosidase inhibitory, and antimicrobial activity of essential oils from *Acacia mollissima* and *Acacia cyclops* cultivated in Tunisia. *Chem Biodivers* 14(10): e1700252.
Juby, K.A., C. Dwivedi, M. Kumar, S. Kota, H.S. Misra, and P.N. Bajaj. 2012. Silver nanoparticle-loaded PVA/gum acacia hydrogel: Synthesis, characterization and antibacterial study. *Carbohydr Polym* 89(3): 906–13.
Kambizi, L., and A.J. Afolayan. 2001. An ethnobotanical study of plants used for the treatment of sexually transmitted diseases (*njovhera*) in Guruve District, Zimbabwe. *J Ethnopharmacol* 77(1): 5–9.
Karim, A.A., and A. Azlan. 2012. Fruit pod extracts as a source of nutraceuticals and pharmaceuticals. *Molecules* 17(10): 11931–46.
Kariuki, A.C., and G.N. Njoroge. 2011. Ethnobotanical and antimicrobial studies of some plants used in Kibwezi (Kenya) for management of lower respiratory tract infections. *Afr J Tradit Complement Altern Med* 8(2): 144–9.
Khalid, M., D. Hassani, M. Bilal, et al. 2017. Identification of oral cavity biofilm forming bacteria and determination of their growth inhibition by *Acacia arabica*, *Tamarix aphylla* L. and *Melia azedarach* L. medicinal plants. *Arch Oral Biol* 81: 175–85.
Khan, R., B. Islam, M. Akram, et al. 2009. Antimicrobial activity of five herbal extracts against multi drug resistant (MDR) strains of bacteria and fungus of clinical origin. *Molecules* 14(2): 586–97.
Kim, G., R.Y. Gan, D. Zhang, et al. 2020. Large-scale screening of 239 traditional Chinese medicinal plant extracts for their antibacterial activities against multidrug-resistant *Staphylococcus aureus* and cytotoxic activities. *Pathogens* 9(3): E185.
Köhler, H.A. 1887. Köhler's Medizinal-Pflanzen in naturgetreuen Abbildungen mit kurz erläuterndem Texte. In: *List of Koehler Images, Public Domain*. Gera-Untermhaus, Germany: Fr. Eugen Köhler. https://commons.wikimedia.org/w/index.php?curid=255255, via Wikimedia Commons.

Krause, D.O., W.J.M. Smith, and C.S. McSweeney. 2004. Use of community genome arrays (CGAs) to assess the effects of *Acacia angustissima* on rumen ecology. *Microbiology* 150(Pt 9): 2899–909.
Kurekci, C., R. Al Jassim, E. Hassan, S.L. Bishop-Hurley, J. Padmanabha, and C.S. McSweeney. 2014. Effects of feeding plant-derived agents on the colonization of *Campylobacter jejuni* in broiler chickens. *Poult Sci* 93(9): 2337–46.
Madureira, A.M., C. Ramalhete, S. Mulhovo, A. Duarte, and M.J. Ferreira. 2012. Antibacterial activity of some African medicinal plants used traditionally against infectious diseases. *Pharm Biol* 50(4): 481–9.
Mahmood, A., A. Mahmood, and R.A. Qureshi. 2012. Antimicrobial activities of three species of family Mimosaceae. *Pak J Pharm Sci* 25(1): 203–6.
Mahmoud, M.F., S.A. Alrumman, and A. el-L. Hesham. 2016. Biological activities of some *Acacia* spp. (Fabaceae) against new clinical isolates identified by ribosomal RNA gene-based phylogenetic analysis. *Pak J Pharm Sci* 29(1): 221–9.
Mandal, P., S.P. Sinha Babu, and N.C. Mandal. 2005. Antimicrobial activity of saponins from *Acacia auriculiformis*. *Fitoterapia* 76(5): 462–5.
Maroyi, A. 2017. *Acacia karroo* Hayne: Ethnomedicinal uses, phytochemistry and pharmacology of an important medicinal plant in Southern Africa. *Asian Pac J Trop Med* 10(4): 351–60.
Mattana, C.M., S.E. Satorres, A. Sosa, M. Fusco, and L.E. Alcará. 2010. Antibacterial activity of extracts of *Acacia aroma* against methicillin-resistant and methicillin-sensitive *Staphylococcus*. *Braz J Microbiol* 41(3): 581–7.
Mazzio, E.A., N. Li, D. Bauer, et al. 2016. Natural product HTP screening for antibacterial (*E. coli 0157:H7*) and anti-inflammatory agents in (LPS from *E. coli* O111:B4) activated macrophages and microglial cells; focus on sepsis. *BMC Complement Altern Med* 16(1): 467.
Mishra, R.K., M. Ramakrishna, V. Mishra, et al. 2016. Pharmaco-phylogenetic investigation of methyl gallate isolated from *Acacia nilotica* (L.) Delile and its cytotoxic effect on NIH3T3 mouse fibroblast. *Curr Pharm Biotechnol* 17(6): 540–8.
Mohammed, A.E., A. Al-Qahtani, A. Al-Mutairi, B. Al-Shamri, and K.F. Aabed. 2018. Antibacterial and cytotoxic potential of biosynthesized silver nanoparticles by some plant extracts. *Nanomaterials (Basel)* 8(6): 382.
Mohieldin, E.A.M., A.M. Muddathir, and T. Mitsunaga. 2017. Inhibitory activities of selected Sudanese medicinal plants on *Porphyromonas gingivalis* and matrix metalloproteinase-9 and isolation of bioactive compounds from *Combretum hartmannianum* (Schweinf) bark. *BMC Complement Altern Med* 17(1): 224.
Mothana, R.A., U. Lindequist, R. Gruenert, and P.J. Bednarski. 2009. Studies of the in vitro anticancer, antimicrobial and antioxidant potentials of selected Yemeni medicinal plants from the island Soqotra. *BMC Complement Altern Med* 9: 7.
Mulaudzi, R.B., A.R. Ndhlala, M.G. Kulkarni, J.F. Finnie, and J. Van Staden. 2011. Antimicrobial properties and phenolic contents of medicinal plants used by the Venda people for conditions related to venereal diseases. *J Ethnopharmacol* 135(2): 330–7.
Mutai, C., C. Bii, C. Vagias, D. Abatis, and V. Roussis. 2009. Antimicrobial activity of *Acacia mellifera* extracts and lupane triterpenes. *J Ethnopharmacol* 123(1): 143–8.
Nalawade, P., P. Mukherjee, and S. Kapoor. 2014. Biosynthesis, characterization and antibacterial studies of silver nanoparticles using pods extract of *Acacia auriculiformis*. *Spectrochim Acta A Mol Biomol Spectrosc* 129: 121–4.
Newbold, C.J., S.M. el Hassan, J. Wang, M.E. Ortega, and R.J. Wallace. 1997. Influence of foliage from African multipurpose trees on activity of rumen protozoa and bacteria. *Br J Nutr* 78(2): 237–49.
Nipanikar, S., S. Chitlange, and D. Nagore. 2017. Evaluation of anti-inflammatory and antimicrobial activity of AHPL/AYCAP/0413 Capsule. *Pharmacognosy Res* 9(3): 273–6.
Ogawa, S., and Y. Yazaki. 2018. Tannins from *Acacia mearnsii* De Wild. Bark: Tannin determination and biological activities. *Molecules* 23(4): 837.
Ojha, D., G. Singh, and Y.N. Upadhyaya. 1969. Clinical evaluation of *Acacia catechu*, Willd. (*Khadira*) in the treatment of lepromatous leprosy. *Int J Lepr Other Mycobact Dis* 37(3): 302–7.
Olajuyigbe, O.O., and A.J. Afolayan. 2012a. *In vitro* antibacterial and time-kill assessment of crude methanolic stem bark extract of *Acacia mearnsii* De Wild against bacteria in shigellosis. *Molecules* 17(2): 2103–18.
Olajuyigbe, O.O., and A.J. Afolayan. 2012b. Pharmacological assessment of the medicinal potential of *Acacia mearnsii* De Wild.: Antimicrobial and toxicity activities. *Int J Mol Sci* 13(4): 4255–67.
Olajuyigbe, O.O., and A.J. Afolayan. 2012c. Synergistic interactions of methanolic extract of *Acacia mearnsii* De Wild. with antibiotics against bacteria of clinical relevance. *Int J Mol Sci* 13(7): 8915–32.
Owayss, A.A., K. Elbanna, J. Iqbal, et al. 2019. In vitro antimicrobial activities of Saudi honeys originating from *Ziziphus spina-christi* L. and *Acacia gerrardii* Benth. trees. *Food Sci Nutr* 8(1): 390–401.

Palombo, E.A., and S.J. Semple. 2002. Antibacterial activity of Australian plant extracts against methicillin-resistant *Staphylococcus aureus* (MRSA) and vancomycin-resistant enterococci (VRE). *J Basic Microbiol* 42(6): 444–8.

Pennacchio, M., A.S. Kemp, R.P. Taylor, K.M. Wickens, and L. Kienow. 2005. Interesting biological activities from plants traditionally used by native Australians. *J Ethnopharmacol* 96(3): 597–601.

Pradeep, A.R., E. Agarwal, P. Bajaj, S.B. Naik, N. Shanbhag, and S.R. Uma. 2012. Clinical and microbiologic effects of commercially available gel and powder containing *Acacia arabica* on gingivitis. *Aust Dent J* 57(3): 312–18.

Quattrocchi, U. 2000. *CRC World Dictionary of Plant Names: Common Names, Scientific Names, Eponyms, Synonyms, and Etymology*. Boca Raton, FL: CRC Press.

Raguvaran, R., B.K. Manuja, M. Chopra, et al. 2017. Sodium alginate and gum acacia hydrogels of ZnO nanoparticles show wound healing effect on fibroblast cells. *Int J Biol Macromol* 96: 185–91.

Rani, P., and N. Khullar. 2004. Antimicrobial evaluation of some medicinal plants for their anti-enteric potential against multi-drug resistant *Salmonella typhi*. *Phytother Res* 18(8): 670–3.

Rao, T.N., Riyazuddin, P. Babji, et al. 2019. Green synthesis and structural classification of *Acacia nilotica* mediated-silver doped titanium oxide (Ag/TiO_2) spherical nanoparticles: Assessment of its antimicrobial and anticancer activity. *Saudi J Biol Sci* 26(7): 1385–91.

Rehman, S., S. Mujtaba Ghauri, and A.N. Sabri. 2016. Impact of plant extracts and antibiotics on biofilm formation of clinical isolates from otitis media. *Jundishapur J Microbiol* 9(1): e29483.

Revathi, S., F.L. Hakkim, N.R. Kumar, et al. 2018. Induction of HT-29 colon cancer cells apoptosis by pyrogallol with growth inhibiting efficacy against drug-resistant *Helicobacter pylori*. *Anticancer Agents Med Chem* 18(13): 1875–84.

Sánchez, E., S. García, and N. Heredia. 2010. Extracts of edible and medicinal plants damage membranes of *Vibrio cholerae*. *Appl Environ Microbiol* 76(20): 6888–94.

Saroya, A.S., and J. Singh. 2020. *Psychoactive Medicinal Plants and Fungal Neurotoxins*. Singapore: Springer Nature Pte.

Sarwar, W. 2016. Pharmacological and phytochemical studies on *Acacia modesta* Wall. A review. *J Phytopharmacol* 5(4): 160–6.

Sharma, A.K., A. Kumar, S.K. Yadav, and A. Rahal. 2014. Studies on antimicrobial and immunomodulatory effects of hot aqueous extract of *Acacia nilotica* L. leaves against common veterinary pathogens. *Vet Med Int* 2014: 747042.

Sherry, S.P. 1968. *The Black Wattle* (Acacia Mearnsii *de Wild*). South Africa: University of Natal.

Sinclair, M. 1932. "Kookaburra," song, © Larrikin Music Publishing Pty Ltd.

Sotohy, S.A., A.N. Sayed, and M.M. Ahmed. 1997. Effect of tannin-rich plant (*Acacia nilotica*) on some nutritional and bacteriological parameters in goats. *Dtsch Tierarztl Wochenschr* 104(10): 432–5.

Tambekar, D.H., B.S. Khante, B.R. Chandak, A.S. Titare, S.S. Boralkar, and S.N. Aghadte. 2009. Screening of antibacterial potentials of some medicinal plants from Melghat forest in India. *Afr J Tradit Complement Altern Med* 6(3): 228–32.

Thakur, M., S. Pandey, A. Mewada, R. Shah, G. Oza, and M. Sharon. 2013. Understanding the stability of silver nanoparticles bio-fabricated using *Acacia arabica* (Babool gum) and its hostile effect on microorganisms. *Spectrochim Acta A Mol Biomol Spectrosc* 109: 344–7.

van Vuuren, S.F., M.N. Nkwanyana, and H. de Wet. 2015. Antimicrobial evaluation of plants used for the treatment of diarrhoea in a rural community in northern Maputaland, KwaZulu-Natal, South Africa. *BMC Complement Altern Med* 15: 53.

Voravuthikunchai, S.P., and S. Limsuwan. 2006. Medicinal plant extracts as anti-*Escherichia coli* O157:H7 agents and their effects on bacterial cell aggregation. *J Food Prot* 69(10): 2336–41.

Voravuthikunchai, S.P., A. Lortheeranuwat, W. Jeeju, T. Sririrak, S. Phongpaichit, and T. Supawita. 2004. Effective medicinal plants against enterohaemorrhagic *Escherichia coli* O157:H7. *J Ethnopharmacol* 94(1): 49–54.

Wu, J., and S. Yu. 2019. Effect of root exudates of *Eucalyptus urophylla* and *Acacia mearnsii* on soil microbes under simulated warming climate conditions. *BMC Microbiol* 19(1): 224.

Yimam, M., L. Brownell, S.G. Do, et al. 2019. Protective effect of UP446 on ligature-induced periodontitis in beagle dogs. *Dent J (Basel)* 7(2): 33.

3 Antifungal Properties of Acacias

This chapter is expected to dovetail with the previous. This is true for several reasons. For one, many of the *in vitro* studies cited in Chapter 2 pertinent to bacteria also included testing of one or more fungi. Thus, some of the references at the end of this chapter are being cited for a second or even third time. However, the content is usually not redundant, and if it is, it should be pardoned since redundancy creates stability and is an important strategy of Nature and of the learning process. Recall Freud's "repetition compulsion" (Chu 1991) and the power of the *mantra* to enact re-*verb*-eration.

A second reason is that some of the mechanisms for suppressing bacteria and fungi overlap. However, there is little overlap, helping to explain fungal resistance in pharmacognostic settings, i.e., in the presence of certain drugs, including plant extracts, where bacteria succumbed. Nevertheless, there is at least some similarity in the survival strategies and in the thwarting of those strategies belonging to bacteria and fungi. Some of the drugs and plant extracts that worked for one may also work for the other.

A third reason is the same as why a survey of antibacterial effects was highly relevant to the initial foray into antiviral effects: because bacteria establish "secondary infections" in the face of viral illness and its often attendant immunological compromise of the host. If this is so, then fungi may create tertiary, opportunistic infections in these very same hosts. The presence of white *Candida* lesions within the mouth of an immune deficiency syndrome patient suffering comorbidly with bacterial pneumonia is an example.

Most of the *Acacia s.l.* species mentioned in relation to the completed research on the antifungal species will have already been profiled in the previous chapters. When they are mentioned for the first time, they will be profiled here.

Finally, this chapter will be considerably shorter than the previous. This does not reflect adversely on the importance of the fungal kingdom, inclusive of medicinal mushrooms, as the source of some of the best and most powerful natural medicines. Rather, it is testament to the awesome power and connectedness of fungi and, in general, their great resistance to drugs, chemicals, and noxious materials.

In an *in vitro* study, essential oils hydro-distilled from the flowers of *Acacia mearnsii* (aka *A. mollissima*) and *A. cyclops* inhibited the growth of a *Candida* sp. (Jelassi et al. 2017). Acetone extract of *A mearnsii*'s stem bark potently inhibited a panel of ten different fungi, but generally at orders of magnitude higher than its effect on both Gram-positive and Gram-negative bacteria. Fungicidal effects (67%) were more prevalent than fungistatic effects (33%). At the doses needed to stop the growth of bacteria, brine shrimp did not die (Olajuyigbe et al. 2012).

Aqueous extracts of the flowers of *Acacia saligna* were found to dose-dependently inhibit mycelial growth on the wood of the chinaberry tree, *Melia azedarach*, when the extracts were applied to its bark. The fungi in question, *Fusarium culmorum*, *Rhizoctonia solani*, and *Penicillium chrysogenum*, isolated from infected *Citrus sinensis*, cause root rot, cankers, and green fruit rot, respectively (Al-Huqail et al. 2019).

An ethyl acetate extract of the bark of *Acacia ataxacantha*, a plant with a reputation around Benin, Africa, as an anti-infective, and a pure compound isolated from the extract, 7-hydroxy-2-methyl-6-[β-galactopyranosyl-propyl]-4H-chromen-4-one, named *acthaside*, significantly inhibited

DOI: 10.1201/9780429440946-3

(a) (b)

FIGURE 3.1 *Coojong, cujong*, the orange, blue-leafed, or golden leaf wattle (*Acacia saligna*), has little sweet nectary glands at the base of its phyllodes, attracting ants that eat leaf-eating insects. Invasive and useful, it is employed for mulch, tanning, landscaping aesthetics, and firewood. (a) *A. saligna*, a twig with flowers, buds, and leaves. (2/21/2016: by Lies Van Rompaey, CC BY 2.0, via Flickr, www.flickr.com/photos/liesvan rompaey/24933214460.) (b) *A. saligna* in bloom. (9/2/2009, Australia: by Sydney Oats, CC BY 2.0, via Flickr, www.flickr.com/photos/57768042@N00/3883792660.)

(a) (b)

FIGURE 3.2 *Melia azedarach* (chinaberry tree). (a) *M. azedarach* flowers, fruits, and leaves. (4/8/2017, East Vista Way, Vista, CA: by K M, CC BY 2.0, via Flickr, www.flickr.com/photos/jccsvq/6700367773.) (b) *M. azedarach* tree trunk. (10/10/2015, Kobe, Hyogo, Japan: by Harum Koh, CC BY 2.0, via Flickr, www.flickr.com/photos/harumkoh/22242835225.)

growth of the common pathogenic yeast, *Candida albicans* (Amoussa, Bourjot et al. 2016). One triterpenoid in the mix, compound 3, was also particularly active against *C. albicans* (Amoussa, Lagnika et al. 2016)

Powdered plant materials from two different acacias, *Vachellia leucophloea* and *V. nilotica*, were extracted successively with hexane, chloroform, acetone, methanol, and water, and the extracts were tested *in vitro* against the fungi *Aspergillus fumigatus*, *A. flavus*, *A. niger*, and *C. albicans*. Polar fractions of *V. leucophloea* and especially of *V. nilotica*, extracted with methanol or water, were efficacious (Dabur et al. 2007).

Seven Sudanese medicinal plants were screened for possible *in vitro* activity against *Madurella mycetomatis*, the etiologic agent underlying eumycetoma, an extremely debilitating chronic infectious fungal disease. Elfadil et al. (2015) found that a methanolic extract of *Vachellia nubica*, as well as those of *Boswellia papyrifera* and *Nigella sativa*, worked well.

FIGURE 3.3 *Citrus sinensis* (cultivated orange) is sensitive to fungi. Photo: *Citrus* x *sinensis* 'Washington' flowers, leaves, and fruit. (4/13/2019, De Luz, CA: by K M, CC BY 2.0, via Flickr, www.flickr.com/photos/131880272@N06/46717321605.)

(a) (b) (c)

FIGURE 3.4 The frost-hardy *Vachellia leucophloea* (*Hta Naung* tree; in Hindi, *reonja*) is extensively used as a folk medicine in Pakistan. The bark is used in preparing alcoholic beverages, and the seeds and their pods are eaten, ground into flour, or, when immature, cooked like a vegetable. (a) *V. leucophloea*, white flowers and foliage (7/20/2014, Thatguni Post, Kanakpura Road, Bengaluru, Karnataka 560082, India: by Ajit Ampalakkad, CC0 1.0, via iNaturalist, www.inaturalist.org/photos/157727902.) (b) *V. leucophloea*, blossoming tree. (10/5/2019, Perambalur, Tamil Nadu, India: by P Jeganathan, CC BY 4.0, via iNaturalist, www.inaturalist.org/photos/56706951.) (c) *V. leucophloea*, seed pods and bark. (1/21/2017, Thatguni Post, Kanakpura Road, Bengaluru, Karnataka 560082, India: by Ajit Ampalakkad, CC0 1.0, via iNaturalist, www.inaturalist.org/photos/157728355.)

FIGURE 3.5 Native to northeast Africa from Egypt to Kenya, *Vachellia nubica* has extensive medical uses among Aboriginal peoples in Kenya and Ethiopia. Photo: *V. nubica* seed pods. (12/11/2016, Selenkay Conservancy, Kenya: by Regina Hart, CC BY 2.0, via Flickr, www.flickr.com/photos/reginahart/33235317521.)

(a)

(b)

FIGURE 3.6 *Vachellia robusta*, "splendid thorn," is a large acacia growing throughout the Afrotropical realm. (a) *V. robusta* tree, habitat. (6/9/2020, Cradle Moon Lakeside Lodge, Muldersdrift, Gauteng, South Africa: by Matthew Fainman, CC BY 4.0, via iNaturalist, www.inaturalist.org/photos/77801763.) (b) *V. robusta* branch with flowers, leaves, and thorns. (9/4/2020, Moloto Rd Roodeplaat, South Africa: by Hildegard Klein, CC BY 4.0, via iNaturalist, https://israel.inaturalist.org/photos/93396911.)

In a survey of 10 Mexican plant species for possible *in vitro* anticandidal action, the most potent against *Candida albicans*, *C. parapsilosis*, *C. tropicalis*, *C. krusei*, and *C. glabrata* was the extract of *Vachellia rigidula* (Garza et al. 2017). Using a large ethnomedical base to generate herbal leads, methanolic extracts of 65 Tanzanian plants were examined for their potential suppression of the following fungi (here in descending order of their sensitivity to the extracts): *Cryptococcus neoformans* > *Candida krusei* > *C. tropicalis* > *C. parapsilosis* >> *C. albicans*, *C. glabrata*. Among the 11 plants whose methanolic extracts were most effective as antifungals were *Vachellia robusta* and *V. nilotica* (Hamza et al. 2006).

Two proteins were isolated from the seeds of *Acacia confusa* that possessed antifungal activity. These included the dimeric *acafusin* with *Rhizoctonia solani* IC(50) of 28 μM (Lam and Ng 2010b) and the chitinase-like *acanonin* with *R. solani* IC(50) of 30 ± 4 μM completely preserved in pH 4–10 and in 0°C–70°C. Acaconin did not inhibit the growth of *Mycosphaerella arachidicola*, *Fusarium oxysporum*, *Helminthosporium maydis*, or *Valsa mali* (Lam and Ng 2010a). Similarly, three novel protease-inhibiting proteins, ApTIA, ApTIB, and ApTIC, were purified from the seeds of *Acacia plumosa*. These *A. plumosa* isoforms inhibited the *in vitro* growth of the fungi *Aspergillus niger*, *Thielaviopsis paradoxa*, and *Colletotrichum* sp. (Lopes et al. 2009).

As an example of the general resistance of fungi, in one *in vitro* study, a methanolic extract of *V. nilotica*, while very effective against a wide range of bacteria, failed to significantly alter the growth of the fungi *Aspergillus niger* or *A. flavus* (Mahmood et al. 2012). However, in another study, sterile filter paper saturated with *A. nilotica* bark extract inhibited the growth of *Candida albicans in vitro*, though not as strongly as that effected by a similar extract of *Punica granatum* pericarp (Pai et al. 2010). A hot water extract of *V. nilotica* leaves exerted significant inhibition of *Aspergillus*

(a) (b)

FIGURE 3.7 *Acacia confusa*, Formosa acacia, a tree growing to 15 m, originated and is still predominantly found in Southeast Asia. Among its ports of debarkation is Hawaii, where it is considered invasive. Its bark is sought by psychonauts for its high content of dimethyltryptamine, and it has uses in the traditional medicine of its region of origin. (a) *A. confusa*, tree and bark. (8/12/2004, Kauhikoa hill, Maui, Hawaii: by Forest & Kim Starr, CC BY 4.0, via Starr Environmental, www.starrenvironmental.com/images/image/?q=24715302945.) (b) *A. confusa*, ripe seeds. (Photo by Steve Hurst, ARS Systematic Botany and Mycology Laboratory @ USDA-NRCS PLANTS Database—USDA, public domain, https://plants.sc.egov.usda.gov/ImageLibrary/original/acco_001_php.jpg.)

FIGURE 3.8 *Acacia plumosa*, "feathery acacia," is native to Brazil. It is sometimes known as *uña de gato*, cat's claw, but care should be taken not to confuse it with the woody liana *Uncaria tomentosa*, also from South America. (Botanical drawing by J. N. Fitch, in W. Curtis, *Curtis's Botanical Magazine* 61 [new ser. 8] [1834]: Plate 3366, in the Biodiversity Heritage Library, public domain, via Flickr, www.flickr.com/photos/biodivlibrary/8203288342.)

(a)

(b)

FIGURE 3.9 *Uncaria tomentosa* (*uña de gato*, cat's claw) is well regarded for its beneficial effect on immunity. (a) *U. tomentosa*, liana with flowers, leaves, and characteristic "claws" (= thorns). (Botanical drawing by R. J. Bazan, in *Annals of the Missouri Botanical Garden* 67 (1980): p. 510, f. 98, via www.plantillustrations.org, public domain, www.plantillustrations.org/illustration.php?id_illustration=275280.) (b) *U. tomentosa* lianas. (By Carlinhaitba, via Shutterstock, www.shutterstock.com/image-photo/uncaria-tomentosa-native-central-south-american-1971968465.)

fumigatus and *A. niger* (Sharma et al. 2014). A methanolic extract of *V. nilotica* bark was the most potent within a group of three leguminous trees, itself, *Prosopis juliflora*, and *Albizia lebbeck*, in inhibiting the crop-destructive soil-borne fungal pathogen, *Sclerotium rolfsii* (Sana et al. 2016). Pure compounds such as *kaempferol*, *apigenin*, and *catechin* isolated from *V. nilotica* were thought to be primarily responsible for the antifungal effect of the extract against wild and fluconazole-resistant strains of *Candida albicans* (Sohaib-Shahzan et al. 2019).

Nevertheless, elsewhere, fungi *Aspergillus ochraceus* and *Curvularia lunata* suffered complete inhibition of conidial germination following exposure to 300 µg/mL or less of a mixture of *acaciasides A* and *B*, acylated bisglycoside saponins isolated from funicles of *Acacia auriculiformis*, whereas to inhibit the growth of bacteria *Bacillus megaterium*, *Salmonella typhimurium*, and *Pseudomonas aeruginosa*, a 700 µg/mL or higher concentration of the mixture was required (Mandal et al. 2005).

The antifungal activity of *Senegalia karroo* is known (Maroyi 2017). In an *in vitro* study of aqueous and methanolic extracts of *S. karroo* bark, activity against *Candida albicans* was recorded (Mulaudzi et al. 2011).

Based on ethnobotanical data in South Africa, *Senegalia caffra* was identified as another acacia with likely antifungal activity, i.e., against *Candida* sp. (Masevhe et al. 2015).

A comparative *in vitro* study of *Acacia mangium vs. auriculiformis* heartwood extracts revealed that the latter suppressed the growth of wood-rotting fungi more effectively than the former. The difference was putatively attributed to higher levels of the flavonoids *3,4′,7,8-tetrahydroxyflavanone* and *teracacidin* found in the heartwood of *A. auriculiformis* as compared to *A. mangium* (Mihara et al. 2005).

(a) (b)

FIGURE 3.10 *Senegalia caffra*, hook-thorn, is a southern African native, prized for its excellent, dense wood for small projects, a favored garden tree, and said to be a source of medicines for Zulu and other indigenous Bantu-speaking African peoples. (a) *S. caffra*, a flowering shrub. (1/24/2018, Stocklands Sappi, KwaZulu-Natal, South Africa: by Peter Warren, CC0 1.0, via iNaturalist, www.inaturalist.org/photos/13111551.) (b) *S. caffra* branches with leaves and ripe seed pods. (5/3/2021, City of Tshwane Metropolitan Municipality, South Africa: by Hildegard Klein, CC BY 4.0, via iNaturalist, www.inaturalist.org/photos/125952818.)

(c) (d)

FIGURE 3.10 (Continued) (c) White *S. caffra* flowers. (9/24/2021, Windy Brow Game Reserve, City of Tshwane Metropolitan Municipality, South Africa: by Matthew Fainman, CC BY 4.0, via iNaturalist, www.inaturalist.org/photos/159341698.) (d) *S. caffra* tree in full bloom. (10/5/2014, Wonderboom, Gauteng, South Africa: by Peter Warren, CC0 1.0, via iNaturalist, www.inaturalist.org/photos/15615608.)

FIGURE 3.11 The straight-trunked *Acacia mangium*, "forest mangrove," grows to a height of 30 m. It is an important commercial source of high-quality timber, and it improves the soil and restores depleted areas. (a) *A. mangium* leaves and flowers. (4/8/2009, Piiholo, Maui, Hawaii: by Forest & Kim Starr, CC BY 4.0, via Starr Environmental, www.starrenvironmental.com/images/image/?q=24924928056.) (b) *A. mangium* forest. (10/15/2009, Piiholo, Maui, Hawaii: by Forest & Kim Starr, CC BY 4.0, via Starr Environmental, www.starrenvironmental.com/images/image/?q=24354631604.)

A methanolic extract of the Socotran tree, *Acacia pennivenia*, efficiently inhibited the *in vitro* growth of *Candida maltosa* (Mothana et al. 2009). Mutai et al. (2009) found that *Candida albicans, C. krusei,* and *Cryptococcus neoformans* were killed by extracts of *Acacia mellifera* made with methanol and dichloromethane.

Honey from *Vachellia gerrardii* showed substantial activity against the dermatophytic fungus *Trichophyton mentagrophytes* (Owayss et al. 2019), while honey from *Senegalia modesta* was active against the fungi *Alternaria alternata* and *Trichoderma harzianum* (Zahoor et al. 2014).

A panel of six human fungal pathogens, namely *Aspergillus fumigatus*, *Absidia corymbifera*, *Candida albicans, C. krusei*, *C. maltosa*, and *Trichophyton mentagrophytes* was used to assess potential antifungal activity from three Yemeni acacias, *Senegalia asak*, *Vachellia nilotica*, and

V. tortilis. The parameter measured was the size of the inhibition zone on the growing plate. (Al-Fatimi et al. 2007). In spite of many challenges due to the inherent resistance of fungi to human interventions of all kinds, the supergenus *Acacia s.l.* offers possible and considerable opportunities for the development of natural, complex antifungal drugs.

REFERENCES

Al-Fatimi, M., M. Wurster, G. Schröder, and U. Lindequist. 2007. Antioxidant, antimicrobial and cytotoxic activities of selected medicinal plants from Yemen. *J Ethnopharmacol* 111(3): 657–66.

Al-Huqail, A.A., S.I. Behiry, M.Z.M. Salem, H.M. Ali, M.H. Siddiqui, and A.Z.M. Salem. 2019. Antifungal, antibacterial, and antioxidant activities of *Acacia Saligna* (Labill.) H. L. Wendl. flower extract: HPLC analysis of phenolic and flavonoid compounds. *Molecules* 24(4): 700.

Amoussa, A.M., M. Bourjot, L. Lagnika, C. Vonthron-Sénécheau, and A. Sanni. 2016. Acthaside: A new chromone derivative from *Acacia ataxacantha* and its biological activities. *BMC Complement Altern Med* 16(1): 506.

Amoussa, A.M., L. Lagnika, M. Bourjot, C. Vonthron-Senecheau, and A. Sanni. 2016. Triterpenoids from *Acacia ataxacantha* DC: Antimicrobial and antioxidant activities. *BMC Complement Altern Med* 16(1): 284.

Chu, J. 1991. The repetition compulsion revisited: Reliving dissociated trauma. *Psychotherapy* 28: 327–32.

Dabur, R., A. Gupta, T.K. Mandal, et al. 2007. Antimicrobial activity of some Indian medicinal plants. *Afr J Tradit Complement Altern Med* 4(3): 313–18.

Elfadil, H., A. Fahal, W. Kloezen, E.M. Ahmed, and W. van de Sande. 2015. The *in vitro* antifungal activity of Sudanese medicinal plants against *Madurella mycetomatis*, the eumycetoma major causative agent. *PLoS Negl Trop Dis* 9(3): e0003488.

Garza, B.A.A., J.L. Arroyo, G.G. González, et al. 2017. Anti-fungal and anti-mycobacterial activity of plants of Nuevo Leon, Mexico. *Pak J Pharm Sci* 30(1): 17–21.

Hamza, O.J., C.J. van den Bout-van den Beukel, M.I. Matee, et al. 2006. Antifungal activity of some Tanzanian plants used traditionally for the treatment of fungal infections. *J Ethnopharmacol* 108(1): 124–32.

Jelassi, A., M. Hassine, M. Besbes Hlila, and H. Ben Jannet. 2017. Chemical composition, antioxidant properties, α-glucosidase inhibitory, and antimicrobial activity of essential oils from *Acacia mollissima* and *Acacia cyclops* cultivated in Tunisia. *Chem Biodivers* 14(10): 10.

Lam, S.K., and T.B. Ng. 2010a. Acaconin, a chitinase-like antifungal protein with cytotoxic and anti-HIV-1 reverse transcriptase activities from *Acacia confusa* seeds. *Acta Biochim Pol* 57(3): 299–304.

Lam, S.K., and T.B. Ng. 2010b. Acafusin, a dimeric antifungal protein from *Acacia confusa* seeds. *Protein Pept Lett* 17(7): 817–22.

Lopes, J.L., N.F. Valadares, D.I. Moraes, J.C. Rosa, H.S. Araújo, and L.M. Beltramini. 2009. Physico-chemical and antifungal properties of protease inhibitors from *Acacia Plumosa*. *Phytochemistry* 70(7): 871–9.

Mahmood, A., A. Mahmood, and R.A. Qureshi. 2012. Antimicrobial activities of three species of family Mimosaceae. *Pak J Pharm Sci* 25(1): 203–6.

Mandal, P., S.P. Sinha Babu, and N.C. Mandal. 2005. Antimicrobial activity of saponins from *Acacia auriculiformis*. *Fitoterapia* 76(5): 462–5.

Maroyi, A. 2017. *Acacia karroo* Hayne: Ethnomedicinal uses, phytochemistry and pharmacology of an important medicinal plant in Southern Africa. *Asian Pac J Trop Med* 10(4): 351–60.

Masevhe, N.A., L.J. McGaw, and J.N. Eloff. 2015. The traditional use of plants to manage candidiasis and related infections in Venda, South Africa. *J Ethnopharmacol* 168: 364–72.

Mihara, R., K.M. Barry, C.L. Mohammed, and T. Mitsunaga. 2005. Comparison of antifungal and antioxidant activities of *Acacia mangium* and *A. auriculiformis* heartwood extracts. *J Chem Ecol* 31(4): 789–804.

Mothana, R.A., U. Lindequist, R. Gruenert, and P.J. Bednarski. 2009. Studies of the in vitro anticancer, antimicrobial and antioxidant potentials of selected Yemeni medicinal plants from the island Soqotra. *BMC Complement Altern Med* 9: 7.

Mulaudzi, R.B., A.R. Ndhlala, M.G. Kulkarni, J.F. Finnie, and J. Van Staden. 2011. Antimicrobial properties and phenolic contents of medicinal plants used by the Venda people for conditions related to venereal diseases. *J Ethnopharmacol* 135(2): 330–7.

Mutai, C., C. Bii, C. Vagias, D. Abatis, and V. Roussis. 2009. Antimicrobial activity of *Acacia mellifera* extracts and lupane triterpenes. *J Ethnopharmacol* 123(1): 143–8.

Olajuyigbe, O.O., and A.J. Afolayan. 2012. Pharmacological assessment of the medicinal potential of *Acacia mearnsii* De Wild.: Antimicrobial and toxicity activities. *Int J Mol Sci* 13(4): 4255–67.

Owayss, A.A., K. Elbanna, J. Iqbal, et al. 2019. In vitro antimicrobial activities of Saudi honeys originating from *Ziziphus spina-christi* L. and *Acacia gerrardii* Benth. trees. *Food Sci Nutr* 8(1): 390–401.
Pai, M.B., G.M. Prashant, K.S. Murlikrishna, K.M. Shivakumar, and G.N. Chandu. 2010. Antifungal efficacy of *Punica granatum, Acacia nilotica, Cuminum cyminum* and *Foeniculum vulgare* on *Candida albicans*: An *in vitro* study. *Indian J Dent Res* 21(3): 334–6.
Sana, N., A. Shoaib, and A. Javaid. 2016. Antifungal potential of leaf extracts of leguminous trees against *Sclerotium Rolfsii. Afr J Tradit Complement Altern Med* 13(5): 54–60.
Sharma, A.K., A. Kumar, S.K. Yadav, and A. Rahal. 2014. Studies on antimicrobial and immunomodulatory effects of hot aqueous extract of *Acacia nilotica* L. leaves against common veterinary pathogens. *Vet Med Int* 2014: 747042.
Sohaib-Shahzan, M., A.S. Smiline Girija, and J. Vijayashree Priyadharsini. 2019. A computational study targeting the mutated L321F of ERG11 gene in *C. albicans*, associated with fluconazole resistance with bioactive compounds from *Acacia nilotica. J Mycol Med* 29(4): 303–9.
Zahoor, M., S. Naz, and M. Sangeen. 2014. Antibacterial, antifungal and antioxidant activities of honey collected from Timergara (Dir, Pakistan). *Pak J Pharm Sci* 27(1): 45–50.

4 Anti-Inflammatory Influence of Acacias

In the Hebrew language, the word for inflammation is *daleket*. The three-consonant "root" of this word is "dlk," pronounced with the vowels as *delek*, which means roughly "fuel." To take it a step further, if you ask for a cigarette light, to "light up," you are said to *le-hadlik*, "to set on fire," to make fire.

In English, the root of the word "inflammation" stands out, too—"flame." Flame results when fuel is "set on fire." When one is excited about something, he or she is "fired up." When prolonged excitement is unchecked, the person may become "burned out."

The organism employs inflammation for a purpose, as a summons and preparation for the immunological battle to fight external germs, viruses, and "bugs" or internal "insurrection," i.e., neoplasia. Chronic, inappropriate, prolonged, non-productive, or "unwanted" inflammation leads to dysfunction and eventually to disease.

Something like inflammation is required to internally process food, i.e., the "digestive fire." Ongoing, inappropriate inflammation of the stomach is gastritis, which can lead to peptic ulcers. As the inflammation progresses, irritable bowel syndrome or Crohn's disease can occur. Something like inflammation occurs also when the body uses insulin to release sugar stores for active burning. But when the pancreatic beta cells fail to respond, the result is chronic inflammation and "insulin resistance." Thus, diabetes and obesity may follow, their etiology powered by chronic inflammation.

Getting "fired up" occurs on the mental level as well, and it can result in internal "resistance" to oneself, psychic fractionation, and, most severely, schizophrenia (Fond et al. 2020; Müller 2018; Pandurangi and Buckley 2020; Upthegrove and Khandaker 2020) or autism (Prata et al. 2017). As fire requires oxygen to burn, oxidative stress is a powerful precipitant of inflammation. The link between oxidative stress and inflammation in schizophrenia (Ma et al. 2020) and autism (Bjørklund et al. 2020; Nadeem et al. 2020) has been elaborated, as has a possible relationship between chronic low-grade inflammation, depression, and dementia (Leonard 2007).

As chronic inflammation provides ground for schizophrenia and autism, so it may also support "somatic psychosis," which is cancer. Cancer and schizophrenia/autism have a lot in common, both in terms of how they make you feel and how they affect you (Bahnson and Bahnson 1969; Brown 2016; Lansky 1982).

Pain is one of the four cardinal signs of inflammation, and thus chronic physical pain is also related to chronic inflammation. This helps explain the prescribing and utility of anti-inflammatory drugs for treating pain. There is a very great overlap between drugs against pain and drugs against inflammation, and sometimes the two are inseparable. However, anti-inflammatory drugs for pain may also themselves inflame and cause new problems, such as the known inflammatory effect of aspirin or steroids on the stomach. Natural drugs that quiet inflammation and/or pain without causing inflammation should confer an advantage both in the clinic and the marketplace. This survey of the potential of the supergenus *Acacia s.l.* to treat inflammation should be seen in this light.

The most prevalent research and application of acacias in medicine, modern as well as traditional, regards the gum of acacia trees, commonly known in pharmacy as "gum arabic." Medicinal acacia gum may come from *Vachellia arabica*, a.k.a. *Vachellia nilotica*, but official "gum arabic" is always from *Senegalia senegal* (*hashab* tree) (Koli et al. 2013), though it is often mixed, substituted, or "adulterated," most commonly with the exudate of *Vachellia seyal* (Ireland et al. 2004; Menzies et al. 1996; Vanloot et al. 2012) or other *Acacia s.l.* species, or even from other genera.

DOI: 10.1201/9780429440946-4

Gum arabic may be used as a matrix for other medications, and it is often included in "controls" or placebos, but the anti-inflammatory properties of gum arabic itself have been documented. For example, in a clinical trial of patients suffering from rheumatoid arthritis, gum arabic decreased inflammatory markers in serum as well as disease severity (Kamal et al. 2018). Also, oral dosing with gum arabic resulted in a significant reduction of C-reactive protein (CRP) in patients with sickle cell anemia (Kaddam and Kaddam 2020). A second clinical trial duplicated this effect of gum arabic on CRP in patients undergoing hemodialysis (Ali et al. 2020). In rabbits, topical application of a gum arabic matrix and ZnO nanoparticles acted intra-synergistically to accelerate scalpel wound healing. Manuja et al. (2020) thought that the gum arabic helped the wound heal by reducing inflammation.

Chemically, gum arabic is a neutral or slightly acidic, branched-chain polysaccharide with repeating *1,3-linked β-D-galactopyranosyl* units comprising both a backbone and two side chains (Ali et al. 2009). Gum arabic decreases inflammation, such as in the case of mice exposed to tobacco water pipe smoke (WPS). The WPS results in cardiotoxicity, for which a compensatory increase in the "master antioxidant switch," nuclear factor erythroid 2-related factor 2 (Nrf2) expression, occurs (Cardozo et al. 2013) and is further potentiated by administration of gum arabic (Nemmar et al. 2019). Gum arabic also protected the intestine and pancreas *in vivo* against inflammatory changes caused by aspirin (Ali et al. 2014; Nasif et al. 2011), the murine liver against acetaminophen (Gamal El-Din et al. 2003), heart against doxorubicin (Abd-Allah et al. 2002), rat kidney against mercuric chloride (Gado et al. 2013) and gentamycin (Al-Majed et al. 2002), rat stomach against ethanol (Taha et al. 2020), and kidney lipids against cis-platinum-induced lipid peroxidation (Al-Majed et al. 2003). In rats, it mitigated surgically induced unilateral obstructive nephropathy (Hammad et al. 2019) and alloxan-induced diabetes (Ahmed et al. 2015). Overall, gum arabic has demonstrated multiple nephroprotective properties, also during acute renal failure, both alone (Ali et al. 2010, 2013, 2014) and in combination with other herbs such as ginger (*Zingiber* sp.) and frankincense (*Boswellia* sp.) (Mahmoud et al. 2012). Gum arabic's nephroprotective effect is mediated by multiple mechanisms, including decreased fibrosis, tubular injury, IL-1β, TNF-α, caspase-3, and MCP-1 ($p < 0.01$); and increased IL-10, antioxidant capacity, and renal CR-1 ($p < 0.001$) (Shafeek et al. 2019).

With selenium-enriched yeast, gum arabic was protective against carbon tetrachloride–induced liver damage in rats (Hamid et al. 2018). Amelioration of intestinal dysfunction *in vivo* by gum arabic was associated with downregulation of the inflammatory mediator NF-kappaB (Wapnir et al. 2008). As a food additive, gum arabic-lycopene nanoparticles protected fat-soluble vitamins from photodegradation (Montenegro et al. 2007). Gum arabic has also been utilized in the fashioning of nanoparticles for broad anti-inflammatory applications, including those containing glycyrrhizin against diabetes (Rani et al. 2017), with ZnO in external dressings for wound healing (Raguvaran et al. 2017), and with curcumin against cancers of the colon/rectum (Udompornmongkol and Chiang 2015), liver (Sarika et al. 2015), or breast (Sarika and Nirmala 2016), and for heart failure (Sunagawa et al. 2012), or myocarditis (Mito et al. 2011). Biomimetic hybrid scaffolds of gum arabic with melatonin facilitated tissue repair, apparently in large part due to the anti-inflammatory effect of the acacia gum (Murali et al. 2016). In a novel formulation, gum arabic was combined with the citrus flavonoid *hesperidin* and silver nanoparticles to create a product aimed at the targeted treatment of rheumatoid arthritis (Rao et al. 2018).

Although gum arabic is Generally Recognized as Safe (GRAS) by the United States Food and Drug Administration (FDA) (Sheu et al. 1986), some questions regarding its safety have been asked and tested. Potential toxicities to and sensitivities of the heart, kidney, and liver have been noted (Bachmann et al. 1978), though it generally emerges that gum arabic is often more protective of such toxicities than causative (Anon 2005; Anderson 1986; Anderson et al. 1982; Collins et al. 1987; Kaddam et al. 2019; Melnick et al. 1983). In one examination of a novel form of gum arabic created by a proprietary process, the toxicity of the material in rats was established for both sexes to be ~3 g/kg (Doi et al. 2006). In another study using typical gum arabic in Sprague Dawley rats, LD_{50} was found to be > 2 g/kg (Schmitt et al. 2008). Allergies to gum arabic have been reported (Turiaf et al. 1959). At least one standardized pill used in Unani medicine is made from gum arabic and is used to treat inflammatory conditions like arthritis or sciatica (Husain et al. 2012).

Fortunately, there is more anti-inflammatory activity within *Acacia s.l.* than gum arabic!

Lectins in the seeds of *Vachellia farnesiana* have been described as blocking inflammatory pathways (Abrantes et al. 2013). A methanolic *V. farnesiana* extract was particularly efficacious in reducing carrageenan-induced rat hind paw edema at 400 mg/kg i.p. (Meckes et al. 2004). Aqueous extracts of *Vachellia karroo* stem bark at 100 or 200 mg/kg *in vivo* significantly reduced carrageenan- or histamine-induced edema and writhes between doses in the acetic acid–induced writhing model and increased reaction time to pain in the tail immersion test, collectively confirming both analgesic and classical anti-inflammatory properties (Adedapo et al. 2008). Similarly, methanolic and ethyl acetate extracts of aerial parts of *Acacia hydaspica* (*A. eburnea*) (Chakrabarty and Maina 2016) at 150 mg/kg significantly inhibited, *in vivo*, yeast-induced pyrexia, pain via acetic acid–induced writhing or the hot plate test, and paw edema in the carrageenan-induced model, thus corroborating anti-inflammatory activity inclusive of inhibiting pyrexia and pain. Several compounds were isolated from the extract and thought to possibly be at least partially responsible for the observed anti-inflammatory effects, including *7-O-galloyl catechin*, *catechin*, and *methyl gallate* (Afsar et al. 2015, 2018b). This group expanded on the anti-inflammatory effect of ethyl acetate, acetone, and other extracts of *A. hydaspica* aerial parts in alleviating *in vivo* damage from cisplatin and/or doxorubicin to the lung (Afsar et al. 2018a), heart (Afsar et al. 2017b, 2019b), and liver (Afsar et al. 2017a, 2019a).

Essential oils were taken from the stem bark and leaves of *Acacia mearnsii*, both when they were dry and when they were still fresh. The essential oil from all samples exerted substantial anti-inflammatory activity in the carrageenan-induced paw edema model. It contained the following most prominent volatile

FIGURE 4.1 *Acacia hydaspica*, a.k.a. *Acacia eburnea*, is common in India, Iran, and parts of China. It has a history of medicinal use among the indigenous people living in its habitat. Photo: *A. hydaspica* J.R. Drumm. ex R. Parker (= *A. eburnea* Willd.), flowers and leaves. (10/4/2010, Eastrenghats, Nellore district, India: by Lalithamba, CC BY 2.0, via Flickr, www.flickr.com/photos/45835639@N04/5050519695.)

compounds in order of their occurrence from highest to lowest concentrations: *cis-verbenol* (30%), *octadecyl alcohol* (26%), and *phytol* (11%) (Avoseh et al. 2015). Tannins from *A. mearnsii* bark have been isolated and encapsulated as a dietary supplement, and extracts have been shown to retard diabetes and obesity, in part through anti-inflammatory mechanisms signaled by the suppression of activity in white adipose tissue of the systemic inflammatory cytokine tumor necrosis factor alpha (TNF-α) (Ogawa and Yazaki 2018). An *A. mearnsii* leaf extract contained 647 mg/kg "gallic acid equivalent" and downregulated expression of anti-inflammatory cytokines IL-1β, COX-2, iNOS, and IL-6 (Xiong et al. 2016), while an *A. mearnsii* bark extract reacted against nitric oxide (NO) and reactive oxygen species (ROS), also suppressing expression of IL-1β, COX-2, iNOS, and IL-6 (Xiong et al. 2017).

High ethanol (70%) hydroethanolic extract was superior to lower ethanol (50%) hydroethanolic extract of young seedless pods of *Vachellia nilotica* and showed superior activity compared to pure ethanol or pure aqueous extracts in inhibiting gastric ulcers in rats induced by pylorus ligation, swimming stress, or nonsteroidal anti-inflammatory drugs (NSAIDs). There was a good correlation between the anti-ulcer activity and the polyphenol content, suggesting that the polyphenols were the active anti-ulcer components in the extracts (Bansal and Goel 2012).

A proprietary combination of *Scutellaria baicalensis* and *Senegalia catechu* was effective in modulating the activities of cyclooxygenase (COX) 1 and 2 and 5-lipoxygenase (LOX) *in vitro*. The 50% inhibitory concentration for the combination was 15 μg/mL for the COX enzymes and 25 μg/mL for the LOX (Burnett et al. 2007). Clinically, it mollified osteoarthritis of the knee while reducing proinflammatory markers (Bitto et al. 2014). This ancient Chinese medical couplet improved lipopolysaccharide-induced lung histopathological changes *in vivo* while inhibiting the release of inflammatory mediators such as TNF-α and IL-1β, in rat bronchoalveolar lavage fluid (Feng et al. 2019). The combination also exerted a "normalization effect" *in vitro* on genes up- or downregulated by a previous exposure to the inflammation-causing lipopolysaccharide (LPS), in the sense that their expression patterns returned to their pre-LPS states (Tseng-Crank et al. 2010). A medicinal tincture and a method for standardizing representative compounds of extracts of *S. catechu* (catechin), *Scutellaria baicalensis* (baicalin), and berberine by using high-performance liquid chromatography (HPLC) were developed and optimized (Kun et al. 2013).

Senegalia catechu (*khadira*) with five other Ayurvedic herbs, including *Azadirachta indica* (*nimba*), *Hemidesmus indicus* (*sariva*), *Rubia cordifolia* (*manjishtha*), and *Solanum nigrum* (*kakmachi*), has been developed for an oral anti-acne product, which also significantly reduced carrageenan-induced rat paw edema, substantiating the anti-inflammatory effect of the mixture (Nipanikar et al. 2017).

A methanolic extract of the dried aqueous extract of *S. catechu* heartwood (*khoyer*) had antihyperglycemic and antinociceptive effects *in vivo* (Rahmatullah et al. 2013). It also protected the mouse lung from the known carcinogen benzo(a)pyrene while downregulating multiple proinflammatory biochemical checkpoints (Shahid et al. 2017). The safety of these extracts, as well as the role of antioxidative processes in creating the anti-inflammatory potency, have been emphasized (Stohs and Bagchi 2015).

The ability of these extracts to reduce TNF-α and NO while increasing IL-10 demonstrates their anti-inflammatory potential. Simultaneously, the *Senegalia catechu/Scutellaria baicalensis* combination enhanced immunocompetence by increasing splenic antibody-producing cells and enhancing the phagocytic responses of peritoneal macrophages (Sunil et al. 2019). The pharmacokinetics of the proprietary UP446—the aforedescribed combination of *S. catechu* and *Scutellaria baicalensis*—with the attention to catechin and epicatechin from *S. catechu* and baicalin and baicalein from *S. baicalensis*—have also been elaborated (Wang et al. 2019). Additional formal studies on toxicology and kinetics have been completed with the UP446 (Yimam et al. 2010, 2013, 2015a, 2015b, 2015c) as well as with the UP1306, which contains *S. catechu* with *Morus alba* instead of *Scutellaria baicalensis* (Yimam et al. 2016, 2018).

A methanolic extract of the leaves of *Acacia modesta* was tested in rodents *in vivo* for possible anti-inflammatory, analgesic, and antiplatelet activities. A single intraperitoneal (i.p.) dose of 50–200 mg/kg significantly reduced carrageenan-induced rat paw edema comparable to that of diclofenac, an NSAID, and dose-dependently (0.5–2.5 mg/mL) reduced arachidonic acid–induced platelet aggregation, an inflammation-mediated symptom. Further, intraperitoneal (i.p.) administration of the methanolic extract (50 and 100 mg/kg) produced significant inhibition (P = 0.01) of acetic acid–induced writhing in mice and increased the pain threshold of mice, which was partially suppressed by the opiate antagonist, *naloxone*, at 5 mg/kg (Bukhari et al. 2010).

A novel compound isolated from the aerial parts of *Vachellia nilotica* was identified as a sex hormone, *3β-acetoxy-17β-hydroxy-androst-5-ene* (CAS 1639–43–6) and shown to exhibit anti-inflammatory properties (Chaubal et al. 2003, 2006). Aqueous extract of *V. nilotica* attenuated carrageenan-induced paw edema and yeast-induced pyrexia in rats and significantly increased hot plate reaction time in mice (Dafallah and al-Mustafa 1996). Catechin derivatives, among other phenolic compounds harvested from *V. nilotica* seed pods, were hypothesized to be responsible for that extract's anti-inflammatory properties (Maldini et al. 2011). A combination of a proanthocyanidin-rich extract (PAC) of *V. nilotica* and a rho iso-alpha acids-rich extract (RIAA) of *Humulus lupulus* (hops) at a RIAA:PAC ratio of 5:1 was found to be optimally effective in exerting anti-inflammatory effects, including inhibition of TNF-α and improvement of insulin-stimulating glucose uptake comparable to that achieved with the conventional agents rosiglitazone and metformin (Tripp et al. 2012).

Chloroformic, hexanic, ketonic, methanolic, methanolic:aqueous, and aqueous extracts obtained from seed pods of *Vachellia farnesiana* were chemically analyzed by the ULPS-ESI-Q-oa/

(a) (b)

FIGURE 4.2 *Senegalia pennata*, "climbing wattle," is better known more as a food than a medicine, its glabrous pinnules prized from early, young, freshly picked shoots and fried as an indigenous vegetable for piquant dishes, curries, chutneys, and pickles in Thailand, India, Burma, Cambodia, Laos, and Indonesia. (a) *Acacia pennata* (L.) Willd. (= *S. pennata* (L.) Maslin) plants. (9/6/2021, Kaeng Ka Am Waterfall, Pha Sawoei, Somdet District, Kalasin 46150, Thailand: by Edd Russell, with permission, via Flickr, www.flickr.com/photos/eddingrid/51433285510/.) (b) *Acacia pennata* (L.) Willd. (= *S. pennata* (L.) Maslin), botanical drawing. (In J. M. Wood and M. S. Evans, *Natal Plants: Descriptions and Figures of Natal Indigenous Plants, with Notes on Their Distribution, Economic Value, Native Names, &c., &c.*, Vol. 3 [Durban: Bennett & Davis, 1902], table 244, via www.plantillustrations.org, public domain, www.plantillustrations.org/illustration.php?id_illustration=192087.)

TOF-MS method and tested *in vitro* against a panel of inflammatory markers and *in vivo* using the ear edema model in CD-1 mice. All extracts significantly reduced ear edema *in vivo* and had an anti-inflammatory effect on the following markers: IL-1β, IL-6, IL-10, TNF-α, and COX. Compounds identified included *methyl gallate*, *gallic acid*, *galloyl glucose isomer 1*, *galloyl glucose isomer 2*, *galloyl glucose isomer 3*, *digalloyl glucose isomer 1*, *digalloyl glucose isomer 2*, *digalloyl glucose isomer 3*, *digalloyl glucose isomer 4*, *hydroxytyrosol acetate*, *quinic acid*, and *caffeoylmalic acid* (Claudia et al. 2018). Earlier, *diosmetin* and *3′,4′,5-trihydroxy-7-methoxyflavone* had been isolated from the roots of *V. farnesiana* and shown to retard superoxide anion generation from leukocytes, and thus they were considered anti-inflammatory, though the novel diterpenes, discovered at the same time from the same plant and the same roots, *acasiane A*, *acasiane B*, *farnesirane A*, and *farnesirane B*, were not reported to have the same effect (Lin et al. 2009). Bronchodilator as well as anti-inflammatory effects of "glycosidal fractions" of *V. farnesiana* have also been reported (Trivedi et al. 1986).

A butanolic extract of the dried leaves of *Senegalia pennata* acted *in vivo* as a chemoprotectant against acetic acid or formalin, increased the threshold of pain sensitivity to pressure, and lessened carrageenan-induced rat paw edema (Dongmo et al. 2005).

An aqueous extract of the seeds of *Acacia saligna* (syn. *cyanophylla*) *in vitro* potentiated the release of superoxide anions from polymorphic leukocytes that had been activated by an induced inflammatory

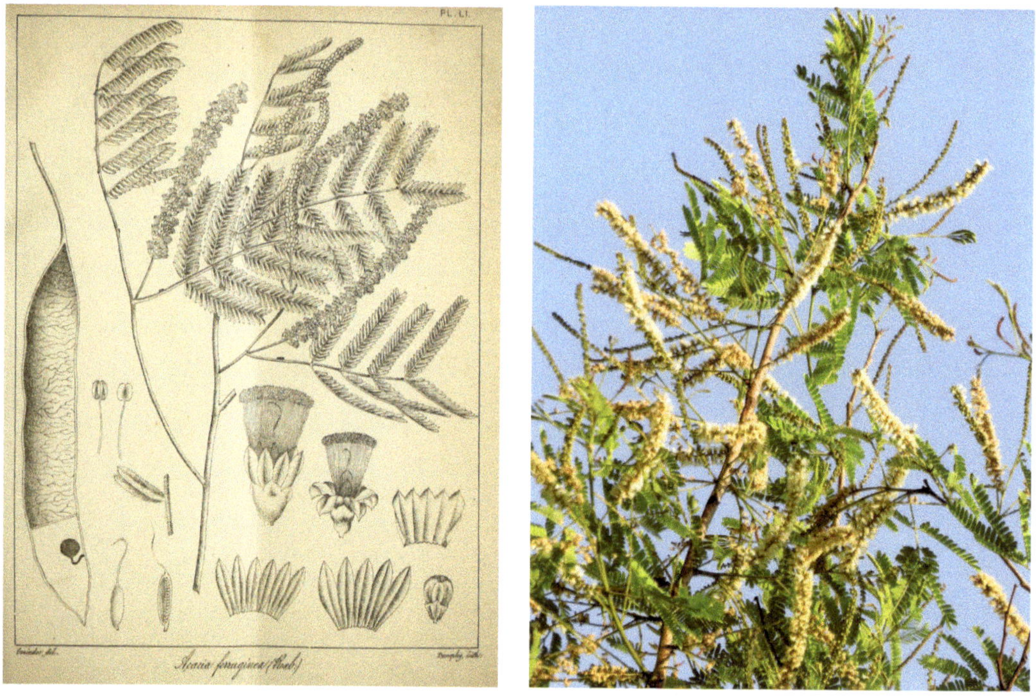

(a) (b)

FIGURE 4.3 *Senegalia ferruginea* is found in Peninsular India and Sri Lanka. There is traditional usage owing to the bitter and astringent properties of the bark for pruritus, leucoderma, ulcers, stomatitis, blood diseases, dysentery, gonorrhea, urinary tract disorders, and diseases of the eye and liver. The bark decoction of *S. ferruginea* is an active ingredient in a commercial preparation for gargling (Faujdar et al. 2018). (a) *Acacia ferruginea* DC. (= *S. ferruginea* (DC.) Pedley), botanical drawing. (By Govindoo, in R. H. Beddome, *Flora sylvatica of Southern India*, Vol. 1 [Madras: Gantz Brothers, 1869], table 51, via www.plantillustrations.org, public domain, www.plantillustrations.org/illustration.php?id_illustration=193086.) (b) *S. ferruginea*, foliage and yellow flowers. (March 2016, Mysore, Karnataka, India: by Shivaprakash, with permission, via iNaturalist, www.inaturalist.org/photos/157311560.)

reaction to calcium pyrophosphate or serum. The superoxide anion potentiation corrected a deficiency of superoxide anions secondary to treatment with the NSAID diclofenac (El Abbouyi et al. 2004). In the essential oils obtained by hydrodistillation from *A. saligna*, the main compound in samples taken from the root was *phenylethyl salicylate* (35%), from the stem, *heptyl valerate* (17%), and from the seed pods, *nonadecane* (36%). The oils taken from *A. saligna* phyllodes, pods, and flowers were potently allelopathic against wild lettuce (*Lactuca virosa*) (El Ayeb-Zakhama et al. 2015).

Hydroalcoholic extracts of *Senegalia ferruginea* bark (Sakthivel and Guruvayoorappan 2016) were tested *in vivo* and found to significantly reduce the inflammatory cytokines TNF-α, interleukin 6 (IL-6), and PGE2, as well as the recto-anal coefficient (RAC = weight of recto-anal tissue (mg)/ body weight (mg)) (Faujdar et al. 2018). *S. ferruginea* extracts were found to inhibit proinflammatory signaling that leads to and sustains cancer (Sakthivel and Guruvayoorappan 2013, 2018), ulcerative colitis (Sakthivel and Guruvayoorappan 2014), and cyclophosphamide-induced immunosuppression and associated urotoxicity (Sakthivel and Guruvayoorappan 2015).

Avicins are pro-apoptotic triterpenoid electrophiles isolated from *Acacia victoriae* that inhibit NF-kappa-B activity (Haridas et al. 2001) with decided effects on increasing the permeability of outer mitochondrial membranes, for example, to allow extrusion of cytochrome C. This permits the compounds to have an overall anti-inflammatory effect. In mouse skin, in response to the "ancient stressor" of ultraviolet light, the following steps were actualized: downregulation of epidermal hyperplasia, attenuated p53 mutation, enhanced apoptosis, less production of 8-hydroxy-2′-deoxyguanosine, and upregulated expression of NADPH:quinone oxidoreductase 1 and heme oxygenase-1. The compounds increase the expression of Nrf2, the ensuing increase in antioxidant protection mechanisms leading to lowered systemic inflammation (Haridas et al. 2004, 2007).

In a cross-cultural study, Rachel W. Li and colleagues at the Australian Center for Complementary Education and Research at the University of Queensland tested ethanolic extracts of 33 traditional herbal medicines from both Aboriginal Australian and Traditional Chinese Medicine for presumed anti-inflammatory effects (based on their traditional uses) *in vitro* against COX-1. All the herbs except for one had some activity, but among the six most potent inhibitors were extracts from the leaves of two acacias, *viz.*, *Acacia ancistrocarpa*, fitzroy wattle, and *Acacia adsurgens*, along with *Ficus racemosa* bark, *Clematis pickeringii* stem, *Tinospora smilacina* stem, and *Morinda citrifolia* fruit powder (Li et al. 2003).

(a) (b)

FIGURE 4.4 *Acacia adsurgens*. First formally described in 1927, this northern Western Australia-endemic shrub may reach a height of 4 m. (a) *A. adsurgens* Maiden & Blakely tree in Australia. (Australia: by M. Fagg, CC-BY 3.0 (Au), in Australian Plant Image Index/Atlas of Living Australia, https://images.ala.org.au/image/details?imageId=47f979b9-052d-4091-81fa-016838a9d6a3.) (b) *A. adsurgens* Maiden & Blakely, branch with leaves and flowers. (7/18/2016, Gibson Desert North WA 0872, Australia: by Graham & Maree Goods, CC-BY 3.0 (Au), in Australian Plant Image Index/Atlas of Living Australia, https://images.ala.org.au/image/details?imageId=d2a7f42b-c6b2-412f-9b73-5a77ed23ebe9.)

Acacia confusa, which is known for its entheogenic properties and was previously discussed regarding its antimicrobial properties, is also known, and has been extensively ethnographically employed in this regard, for treating a wide range of inflammation-related disorders (Lin et al. 2018). In a study of *A. confusa*'s heartwood, an ethanolic extract was prepared and further fractionated. The ethyl acetate (EtOAc) fraction had the highest activity in suppressing NO production from lipopolysaccharide-treated RAW 264.7 macrophages, and it yielded the following novel compounds: *7,8,3′,4′-tetrahydroxy-4-methoxyflavan-3-ol*, *7,8,3′,4′-tetrahydroxyflavone*, *7,8,3′-trihydroxy-3,4′-dimethoxyflavone*, *7,3′,4′-trihydroxyflavone*, and *7,3′,4′-trihydroxy-3-methoxyflavone*. Also from the EtOAc fraction, *melanoxetin* (*3,7,8,3′,4′-pentahydroxyflavone*) suppressed LPS-induced NO and prostaglandin E 2 (PGE 2) production and completely suppressed gene expression of inducible NO synthase (iNOS) and cyclooxygenase-2 (COX-2) at 50 and 100 μM, respectively (Wu et al. 2008).

Maldini et al. (2009) found that a chloroformic extract of the bullhorn wattle, *Acacia cornigera*, significantly retarded ear tip edema in mice that had been exposed to croton oil. This was especially so when the extract was taken from the leaves.

Aqueous and organic solvents were used to extract *Senegalia visco*. The extracts were tested in rats and found nontoxic. Further, they suppressed carrageenan-induced ear edema, phospholipase-A2 edema, and cotton pellet–induced granuloma, revealing in part the scope of the extracts' anti-inflammatory effects. Notable compounds in the extracts partially responsible for the extracts' anti-inflammatory actions included the triterpenoids lupeol, α-amyrin, and β-amyrin (Pedernera et al. 2010).

(a) (b)

FIGURE 4.5 *Senegalia visco*. Visco is native to South America, and it may grow—in Bolivia, at 1500 to 3000 m altitude—to individual tree heights of 10 m or more. (a) *S. visco* tree in Argentina. (11/23/2019, Luján de Cuyo, Mendoza, Argentina: by J. Benjamin Bender, with permission, via iNaturalist, www.inaturalist.org/photos/56825086.) (b) *S. visco*, leaves and yellow flowers. (11/23/2019, Luján de Cuyo, Mendoza, Argentina: by J. Benjamin Bender, with permission, via iNaturalist, www.inaturalist.org/photos/56825091.)

REFERENCES

Abd-Allah, A.R., A.A. Al-Majed, A.M. Mostafa, O.A. Al-Shabanah, A.G. Din, and M.N. Nagi. 2002. Protective effect of arabic gum against cardiotoxicity induced by doxorubicin in mice: A possible mechanism of protection. *J Biochem Mol Toxicol* 16(5): 254–9.

Abrantes, V.E., B.A. Matias da Rocha, R. Batista da Nóbrega, et al. 2013. Molecular modeling of lectin-like protein from *Acacia farnesiana* reveals a possible anti-inflammatory mechanism in carrageenan-induced inflammation. *Biomed Res Int* 2013: 253483.

Adedapo, A.A., M.O. Sofidiya, P.J. Masika, and A.J. Afolayan. 2008. Anti-inflammatory and analgesic activities of the aqueous extract of *Acacia karroo* stem bark in experimental animals. *Basic Clin Pharmacol Toxicol* 103(5): 397–400.

Afsar, T., M.R. Khan, S. Razak, S. Ullah, and B. Mirza. 2015. Antipyretic, anti-inflammatory and analgesic activity of *Acacia hydaspica* R. Parker and its phytochemical analysis. *BMC Complement Altern Med* 15: 136.

Afsar, T., S. Razak, and A. Almajwal. 2019a. Effect of *Acacia hydaspica* R. Parker extract on lipid peroxidation, antioxidant status, liver function test and histopathology in doxorubicin treated rats. *Lipids Health Dis* 18(1): 126.

Afsar, T., S. Razak, A. Almajwal, and M.R. Khan. 2018a. *Acacia hydaspica* R. Parker ameliorates cisplatin induced oxidative stress, DNA damage and morphological alterations in rat pulmonary tissue [correction published in 2019 in *BMC Complement Altern Med* 19(1): 302]. *BMC Complement Altern Med* 18(1): 49.

Afsar, T., S. Razak, A. Almajwal, and M.R. Khan. 2017a. Modulatory influence of *Acacia hydaspica* R. Parker ethyl acetate extract against cisplatin inveigled hepatic injury and dyslipidemia in rats. *BMC Complement Altern Med* 17(1): 307.

Afsar, T., S. Razak, A. Almajwal, M. Shabbir, and M.R. Khan. 2019b. Evaluating the protective potency of *Acacia hydaspica* R. Parker on histological and biochemical changes induced by Cisplatin in the cardiac tissue of rats. *BMC Complement Altern Med* 19(1): 182.

Afsar, T., S. Razak, K.M. Batoo, and M.R. Khan. 2017b. *Acacia hydaspica* R. Parker prevents doxorubicin-induced cardiac injury by attenuation of oxidative stress and structural cardiomyocyte alterations in rats. *BMC Complement Altern Med* 17(1): 554.

Afsar, T., S. Razak, M. Shabbir, and M.R. Khan. 2018b. Antioxidant activity of polyphenolic compounds isolated from ethyl-acetate fraction of *Acacia hydaspica* R. Parker. *Chem Cent J* 12(1): 5.

Ahmed, A.A., J.S. Fedail, H.H. Musa, A.A. Kamboh, A.Z. Sifaldin, and T.H. Musa. 2015. Gum Arabic extracts protect against hepatic oxidative stress in alloxan induced diabetes in rats. *Pathophysiology* 22(4): 189–94.

Al-Majed, A.A., A.M. Mostafa, A.C. Al-Rikabi, and O.A. Al-Shabanah. 2002. Protective effects of oral arabic gum administration on gentamicin-induced nephrotoxicity in rats. *Pharmacol Res* 46(5): 445–51.

Al-Majed, A.A., A.R. Abd-Allah, A.C. Al-Rikabi, O.A. Al-Shabanah, and A.M. Mostafa. 2003. Effect of oral administration of Arabic gum on cisplatin-induced nephrotoxicity in rats. *J Biochem Mol Toxicol* 17(3): 146–53.

Ali, B.H., M. Al Za'abi, A. Ramkumar, J. Yasin, and A. Nemmar. 2014. Anemia in adenine-induced chronic renal failure and the influence of treatment with gum acacia thereon. *Physiol Res* 63(3): 351–8.

Ali, B.H., S. Al-Salam, I. Al Husseni, et al. 2010. Effects of Gum arabic in rats with adenine-induced chronic renal failure. *Exp Biol Med (Maywood)* 235(3): 373–82.

Ali, B.H., S. Beegam, I. Al-Lawati, M.I. Waly, M. Al Za'abi, and A. Nemmar. 2013. Comparative efficacy of three brands of gum acacia on adenine-induced chronic renal failure in rats. *Physiol Res* 62(1): 47–56.

Ali, B.H., A. Ziada, and G. Blunden. 2009. Biological effects of gum arabic: A review of some recent research. *Food Chem Toxicol* 47(1): 1–8.

Ali, N.E., L.A. Kaddam, S.Y. Alkarib, et al. 2020. Gum arabic (*Acacia Senegal*) augmented total antioxidant capacity and reduced c-reactive protein among haemodialysis patients in phase II trial. *Int J Nephrol* 2020: 7214673.

Anderson, D.M. 1986. Evidence for the safety of gum arabic (*Acacia senegal* (L.) Willd.) as a food additive: A brief review. *Food Addit Contam* 3(3): 225–30.

Anderson, D.M., P. Ashby, A. Busuttil, et al. 1982. Subchronic effects of gum arabic (*Acacia*) in the rat. *Toxicol Lett* 14(3–4): 221–7.

Avoseh, O.N., O.O. Oyedeji, K. Aremu, et al. 2015. Chemical composition and anti-inflammatory activities of the essential oils from *Acacia mearnsii* de Wild. *Nat Prod Res* 29(12): 1184–8.

Bachmann, E., E. Weber, M. Post, and G. Zbinden. 1978. Biochemical effects of gum arabic, gum tragacanth, methylcellulose and carboxymethylcellulose-Na in rat heart and liver. *Pharmacology* 17(1): 39–49.

Bahnson, M.B., and C.B. Bahnson. 1969. Ego defenses in cancer patients. *Ann NY Acad Sci* 164(2): 546–59.

Bansal, V.K., and R.K. Goel. 2012. Gastroprotective effect of *Acacia nilotica* young seedless pod extract: Role of polyphenolic constituents. *Asian Pac J Trop Med* 5(7): 523–8.

Bitto, A., F. Squadrito, N. Irrera, et al. 2014. Flavocoxid, a nutraceutical approach to blunt inflammatory conditions. *Mediators Inflamm* 2014: 790851.

Bjørklund, G., N.A. Meguid, M.A. El-Bana, et al. 2020. Oxidative stress in autism spectrum disorder. *Mol Neurobiol* 57(5): 2314–32.

Brown, J.S., Jr. 2016. Cancer immune equilibrium and schizophrenia have similar interferon-γ, tumor necrosis factor-α, and interleukin expression: A tumor model of schizophrenia. *Schizophr Bull* 42(6): 1407–17.

Bukhari, I.A., R.A. Khan, A.H. Gilani, S. Ahmed, and S.A. Saeed. 2010. Analgesic, anti-inflammatory and anti-platelet activities of the methanolic extract of *Acacia modesta* leaves. *Inflammopharmacology* 18(4): 187–96.

Burnett, B.P., Q. Jia, Y. Zhao, and R.M. Levy. 2007. A medicinal extract of *Scutellaria baicalensis* and *Acacia catechu* acts as a dual inhibitor of cyclooxygenase and 5-lipoxygenase to reduce inflammation. *J Med Food* 10(3): 442–51.

Cardozo, L.F., L.M. Pedruzzi, P. Stenvinkel, et al. 2013. Nutritional strategies to modulate inflammation and oxidative stress pathways via activation of the master antioxidant switch Nrf2. *Biochimie* 95(8): 1525–33.

Chakrabarty, T., and V. Maina. 2016. *Vachellia hydaspica* and *V. pseudoeburnea* (Fabaceae: Mimosoideae) synonymized under *V. eburnea*. *Phytotaxa* 257(3): 297.

Chaubal, R., A.M. Mujumdar, A. Misar, V.H. Deshpande, and N.R. Deshpande. 2006. Structure-activity relationship study of androstene steroids with respect to local anti-inflammatory activity. *Arzneimittelforschung* 56(6): 394–8.

Chaubal, R., A.M. Mujumdar, V.G. Puranik, V.H. Deshpande, and N.R. Deshpande. 2003. Isolation and X-ray study of an anti-inflammatory active androstene steroid from *Acacia nilotica*. *Planta Med* 69(3): 287–8.

Claudia, D.P., C.H. Mario, N.O. Arturo, et al. 2018. Phenolic compounds in organic and aqueous extracts from *Acacia farnesiana* pods analyzed by ULPS-ESI-Q-oa/TOF-MS. In vitro antioxidant activity and anti-inflammatory response in CD-1 mice. *Molecules* 23(9): 2386.

Collins, T.F., J.J. Welsh, T.N. Black, S.L. Graham, and L.H. Brown. 1987. Study of the teratogenic potential of gum arabic. *Food Chem Toxicol* 25(11): 815–21.

Dafallah, A.A., and Z. al-Mustafa. 1996. Investigation of the anti-inflammatory activity of *Acacia nilotica* and *Hibiscus sabdariffa*. *Am J Chin Med* 24(3–4): 263–9.

Doi, Y., T. Ichihara, A. Hagiwara, et al. 2006. A ninety-day oral toxicity study of a new type of processed gum Arabic, from Acacia tree (*Acacia Senegal*) exudates, in F344 rats. *Food Chem Toxicol* 44(4): 560–6.

Dongmo, A.B., T. Nguelefack, and M.A. Lacaille-Dubois. 2005. Antinociceptive and anti-inflammatory activities of *Acacia pennata* wild (Mimosaceae). *J Ethnopharmacol* 98(1–2): 201–6.

El Abbouyi, A., M. Toumi, Y. El Hachimi, and A. Jossang. 2004. In vitro effects of aqueous seeds extract of *Acacia cyanophylla* on the opsonized zymosan-induced superoxide anions production by rat polymorphonuclear leukocytes. *J Ethnopharmacol* 91(1): 159–65.

El Ayeb-Zakhama, A., L. Sakka-Rouis, A. Bergaoui, G. Flamini, H. Ben Jannet, and F. Harzallah-Skhiri. 2015. Chemical composition and allelopathic potential of essential oils obtained from *Acacia cyanophylla* Lindl. cultivated in Tunisia. *Chem Biodivers* 12(4): 615–26.

Faujdar, S., B. Sati, S. Sharma, A.K. Pathak, and S.K. Paliwal. 2018. Phytochemical evaluation and anti-hemorrhoidal activity of bark of *Acacia ferruginea* DC. *J Tradit Complement Med* 9(2): 85–9.

Feng, T., L. Zhou, S. Gai, et al. 2019. *Acacia catechu* (L.f.) Willd and *Scutellaria baicalensis* Georgi extracts suppress LPS-induced pro-inflammatory responses through NF-κB, MAPK, and PI3K-Akt signaling pathways in alveolar epithelial type II cells. *Phytother Res* 33(12): 3251–60.

Fond, G., C. Lançon, T. Korchia, P. Auquier, and L. Boyer. 2020. The role of inflammation in the treatment of schizophrenia. *Front Psychiatry* 11: 160.

Gado, A.M., and B.A. Aldahmash. 2013. Antioxidant effect of arabic gum against mercuric chloride-induced nephrotoxicity. *Drug Des Devel Ther* 7: 1245–52.

Gamal El-Din, A.M., A.M. Mostafa, O.A. Al-Shabanah, A.M. Al-Bekairi, and M.N. Nagi. 2003. Protective effect of arabic gum against acetaminophen-induced hepatotoxicity in mice. *Pharmacol Res* 48(6): 631–5.

Hamid, M., Y. Abdulrahim, D. Liu, G. Qian, A. Khan, and K. Huang. 2018. The hepatoprotective effect of selenium-enriched yeast and gum arabic combination on carbon tetrachloride-induced chronic liver injury in rats. *J Food Sci* 83(2): 525–34.

Hammad, F.T., S.A. Salam, A. Nemmar, M. Ali, and L. Lubbad. 2019. The effect of arabic gum on renal function in reversible unilateral ureteric obstruction. *Biomolecules* 9(1): 25.

Haridas, V., C.J. Arntzen, and J.U. Gutterman. 2001. Avicins, a family of triterpenoid saponins from *Acacia victoriae* (Bentham), inhibit activation of nuclear factor-kappaB by inhibiting both its nuclear localization and ability to bind DNA. *Proc Natl Acad Sci U S A* 98(20): 11557–62.
Haridas, V., M. Hanausek, G. Nishimura, et al. 2004. Triterpenoid electrophiles (avicins) activate the innate stress response by redox regulation of a gene battery. *J Clin Invest* 113(1): 65–73.
Haridas, V., X. Li, T. Mizumachi, et al. 2007. Avicins, a novel plant-derived metabolite lowers energy metabolism in tumor cells by targeting the outer mitochondrial membrane. *Mitochondrion* 7(3): 234–40.
Husain, S.F., I. Ahmad, and S. Shamsi. 2012. Standardization of the finished product: *Habbe Irqun Nisa*: A Unani anti-inflammatory formulation. *Anc Sci Life* 32(1): 38–44.
Ireland, H.E., A. Clutterbuck, J.P. Cloquet, et al. 2004. The development of immunoassays to identify and quantify species source of gum Arabic. *J Agric Food Chem* 52(26): 7804–8.
Kaddam, L.A., I. Fdl-Elmula, O.A. Eisawi, H.A. Abdelrazig, M.K. Elnimeiri, and A.M. Saeed. 2019. Biochemical effects and safety of gum arabic (*Acacia Senegal*) supplementation in patients with sickle cell anemia. *Blood Res* 54(1): 31–7.
Kaddam, L.A., and A.S. Kaddam. 2020. Effect of Gum Arabic (*Acacia senegal*) on C-reactive protein level among sickle cell anemia patients. *BMC Res Notes* 13(1): 162.
Kamal, E., L.A. Kaddam, M. Dahawi, et al. 2018. Gum Arabic fibers decreased inflammatory markers and disease severity score among rheumatoid arthritis patients, phase II trial. *Int J Rheumatol* 2018: 4197537.
Koli, A.O., A.M. Eltayeb, E.M. Sanjak, and M.H. Mohammed. 2013. Socio-economic aspects of Gum Arabic production in Dalanj area, South Korodofan, Sudan. *Pak J Biol Sci* 16(21): 1407–10.
Kun, L., W. Fei, D. Yun-Peng, P. Xiao-Qiu, E. Yun-Juan, and Z. Huo-Li. 2013. Simultaneous determination of contents of three active components in Jiejia Tincture by HPLC method. *Afr J Tradit Complement Altern Med* 10(5): 370–4.
Lansky, P. 1982. Possibility of hypnosis as an aid in cancer therapy. *Perspect Biol Med* 25(3): 496–503.
Leonard, B.E. 2007. Inflammation, depression and dementia: Are they connected? *Neurochem Res* 32(10): 1749–56.
Li, R.W., S.P. Myers, D.N. Leach, G.D. Lin, and G. Leach. 2003. A cross-cultural study: Anti-inflammatory activity of Australian and Chinese plants. *J Ethnopharmacol* 85(1): 25–32.
Lin, A.S., C.R. Lin, Y.C. Du, et al. 2009. Acasiane A and B and farnesirane A and B, diterpene derivatives from the roots of *Acacia farnesiana*. *Planta Med* 75(3): 256–61.
Lin, H.Y., T.C. Chang, and S.T. Chang. 2018. A review of antioxidant and pharmacological properties of phenolic compounds in *Acacia confusa*. *J Tradit Complement Med* 8(4): 443–50.
Ma, J., L. Yan, T. Guo, et al. 2020. A pilot study of biomarkers of oxidative stress in serum and schizophrenia. *Psychiatry Res* 284: 112757.
Mahmoud, M.F., A.A. Diaai, and F. Ahmed. 2012. Evaluation of the efficacy of ginger, Arabic gum, and *Boswellia* in acute and chronic renal failure. *Ren Fail* 34(1): 73–82.
Maldini, M., P. Montoro, A.I. Hamed, et al. 2011. Strong antioxidant phenolics from *Acacia nilotica*: Profiling by ESI-MS and qualitative-quantitative determination by LC-ESI-MS. *J Pharm Biomed Anal* 56(2): 228–39.
Maldini, M., S. Sosa, P. Montoro, et al. 2009. Screening of the topical anti-inflammatory activity of the bark of *Acacia cornigera* Willdenow, *Byrsonima crassifolia* Kunth, *Sweetia panamensis* Yakovlev and the leaves of *Sphagneticola trilobata* Hitchcock. *J Ethnopharmacol* 122(3): 430–3.
Manuja, A., R. Raguvaran, B. Kumar, A. Kalia, and B.N. Tripathi. 2020. Accelerated healing of full thickness excised skin wound in rabbits using single application of alginate/acacia based nanocomposites of ZnO nanoparticles. *Int J Biol Macromol* 155: 823–833.
Meckes, M., A.D. David-Rivera, V. Nava-Aguilar, and A. Jimenez. 2004. Activity of some Mexican medicinal plant extracts on carrageenan-induced rat paw edema. *Phytomedicine* 11(5): 446–51.
Melnick, R.L., J. Huff, J.K. Haseman, et al. 1983. Chronic effects of agar, guar gum, gum arabic, locust-bean gum, or Tara gum in F344 rats and B6C3F1 mice. *Food Chem Toxicol* 21(3): 305–11.
Menzies, A.R., M.E. Osman, A.A. Malik, and T.C. Baldwin. 1996. A comparison of the physicochemical and immunological properties of the plant gum exudates of *Acacia senegal* (gum arabic) and *Acacia seyal* (gum tahla). *Food Addit Contam* 13(8): 991–9.
Mito, S., R.A. Thandavarayan, M. Ma, et al. 2011. Inhibition of cardiac oxidative and endoplasmic reticulum stress-mediated apoptosis by curcumin treatment contributes to protection against acute myocarditis. *Free Radic Res* 45(10): 1223–31.
Montenegro, M.A., I.L. Nunes, A.Z. Mercadante, and C.D. Borsarelli. 2007. Photoprotection of vitamins in skimmed milk by an aqueous soluble lycopene-gum arabic microcapsule. *J Agric Food Chem* 55(2): 323–9.

Müller, N. 2018. Inflammation in schizophrenia: Pathogenetic aspects and therapeutic considerations. *Schizophr Bull* 44(5): 973–82.
Murali, R., P. Thanikaivelan, and K. Cheirmadurai. 2016. Melatonin in functionalized biomimetic constructs promotes rapid tissue regeneration in Wistar albino rats. *J Mater Chem B* 4(35): 5850–62.
Nadeem, A., S.F. Ahmad, L.Y. Al-Ayadhi, et al. 2020. Differential regulation of Nrf2 is linked to elevated inflammation and nitrative stress in monocytes of children with autism. *Psychoneuroendocrinology* 113: 104554.
Nasif, W.A., M. Lotfy, and M.R. Mahmoud. 2011. Protective effect of gum acacia against the aspirin induced intestinal and pancreatic alterations. *Eur Rev Med Pharmacol Sci* 15(3): 285–92.
Nemmar, A., S. Al-Salam, S. Beegam, P. Yuvaraju, and B.H. Ali. 2019. Gum Arabic ameliorates impaired coagulation and cardiotoxicity induced by water-pipe smoke exposure in mice. *Front Physiol* 10: 53.
Nipanikar, S., S. Chitlange, and D. Nagore. 2017. Evaluation of anti-inflammatory and antimicrobial activity of AHPL/AYCAP/0413 capsule. *Pharmacognosy Res* 9(3): 273–6.
Ogawa, S., and Y. Yazaki. 2018. Tannins from *Acacia mearnsii* De Wild. Bark: Tannin determination and biological activities. *Molecules* 23(4): 837.
Pandurangi, A.K., and P.F. Buckley. 2020. Inflammation, antipsychotic drugs, and evidence for effectiveness of anti-inflammatory agents in schizophrenia. *Curr Top Behav Neurosci* 44: 227–44.
Pedernera, A.M., T. Guardia, C.E. Calderón, et al. 2010. Anti-inflammatory effect of *Acacia visco* extracts in animal models. *Inflammopharmacology* 18(5): 253–60.
Prata, J., S.G. Santos, M.I. Almeida, R. Coelho, and M.A. Barbosa. 2017. Bridging autism spectrum disorders and schizophrenia through inflammation and biomarkers: Pre-clinical and clinical investigations. *J Neuroinflammation* 14(1): 179.
Raguvaran, R., B.K. Manuja, M. Chopra, et al. 2017. Sodium alginate and gum acacia hydrogels of ZnO nanoparticles show wound healing effect on fibroblast cells. *Int J Biol Macromol* 96: 185–91.
Rahmatullah, M., M. Hossain, A. Mahmud, et al. 2013. Antihyperglycemic and antinociceptive activity evaluation of "khoyer" prepared from boiling the wood of *Acacia catechu* in water. *Afr J Tradit Complement Altern Med* 10(4): 1–5.
Rani, R., S. Dahiya, D. Dhingra, N. Dilbaghi, K.H. Kim, and S. Kumar. 2017. Evaluation of anti-diabetic activity of glycyrrhizin-loaded nanoparticles in nicotinamide-streptozotocin-induced diabetic rats. *Eur J Pharm Sci* 106: 220–30.
Rao, K., S. Aziz, T. Roome, et al. 2018. Gum acacia stabilized silver nanoparticles based nano-cargo for enhanced anti-arthritic potentials of hesperidin in adjuvant induced arthritic rats. *Artif Cells Nanomed Biotechnol* 46(Suppl. 1): 597–607.
Sakthivel, K.M., and C. Guruvayoorappan. 2013. *Acacia ferruginea* inhibits tumor progression by regulating inflammatory mediators—(TNF-α, iNOS, COX-2, IL-1β, IL-6, IFN-γ, IL-2, GM-CSF) and pro-angiogenic growth factor—VEGF. *Asian Pac J Cancer Prev* 14(6): 3909–19.
Sakthivel, K.M., and C. Guruvayoorappan. 2014. Protective effect of *Acacia ferruginea* against ulcerative colitis via modulating inflammatory mediators, cytokine profile and NF-κB signal transduction pathways. *J Environ Pathol Toxicol Oncol* 33(2): 83–98.
Sakthivel, K.M., and C. Guruvayoorappan. 2015. *Acacia ferruginea* inhibits cyclophosphamide-induced immunosuppression and urotoxicity by modulating cytokines in mice. *J Immunotoxicol* 12(2): 154–63.
Sakthivel, K.M., and C. Guruvayoorappan. 2016. *Acacia ferruginea* inhibits inflammation by regulating inflammatory iNOS and COX-2. *J Immunotoxicol* 13(1): 127–35.
Sakthivel, K.M., and C. Guruvayoorappan. 2018. Targeted inhibition of tumor survival, metastasis and angiogenesis by *Acacia ferruginea* mediated regulation of VEGF, inflammatory mediators, cytokine profile and inhibition of transcription factor activation. *Regul Toxicol Pharmacol* 95: 400–11.
Sarika, P.R., N.R. James, P.R. Kumar, D.K. Raj, and T.V. Kumary. 2015. Gum Arabic-curcumin conjugate micelles with enhanced loading for curcumin delivery to hepatocarcinoma cells. *Carbohydr Polym* 134: 167–74.
Sarika, P.R., and R.J. Nirmala. 2016. Curcumin loaded gum arabic aldehyde-gelatin nanogels for breast cancer therapy. *Mater Sci Eng C Mater Biol Appl* 65: 331–7.
Schmitt, D., N. Tran, S. Riefler, et al. 2008. Toxicologic evaluation of modified gum acacia: Mutagenicity, acute and subchronic toxicity. *Food Chem Toxicol* 46(3): 1048–54.
Shafeek, F., N. Abu-Elsaad, A. El-Karef, and T. Ibrahim. 2019. Gum Acacia mitigates diclofenac nephrotoxicity by targeting monocyte chemoattractant protein-1, complement receptor-1 and pro-apoptotic pathways. *Food Chem Toxicol* 129: 162–8.
Shahid, A., R. Ali, N. Ali, et al. 2017. Methanolic bark extract of *Acacia catechu* ameliorates benzo(a)pyrene induced lung toxicity by abrogation of oxidative stress, inflammation, and apoptosis in mice. *Environ Toxicol* 32(5): 1566–77.

Sheu, C.W., K.T. Cain, C.J. Rushbrook, T.A. Jorgenson, and W.M. Generoso. 1986. Tests for mutagenic effects of ammoniated glycyrrhizin, butylated hydroxytoluene, and gum arabic in rodent germ cells. *Environ Mutagen* 8(3): 357–67.

Stohs, S.J., and D. Bagchi. 2015. Antioxidant, anti-inflammatory, and chemoprotective properties of *Acacia catechu* heartwood extracts. *Phytother Res* 29(6): 818–24.

Sunagawa, Y., H. Wada, H. Suzuki, et al. 2012. A novel drug delivery system of oral curcumin markedly improves efficacy of treatment for heart failure after myocardial infarction in rats. *Biol Pharm Bull* 35(2): 139–44.

Sunil, M.A., V.S. Sunitha, E.K. Radhakrishnan, and M. Jyothis. 2019. Immunomodulatory activities of *Acacia catechu*, a traditional thirst quencher of South India. *J Ayurveda Integr Med* 10(3): 185–91.

Taha, M.S., E.M. El-Sherbiny, and H.F. Osman. 2020. Anti-ulcerogenic activity of Arabic Gum in gastric mucosal injury induced by ethanol in male albino rats. *Appl Physiol Nutr Metab* 45(7): 731–6.

Tripp, M.L., G. Darland, V.R. Konda, et al. 2012. Optimized mixture of hops rho iso-alpha acids-rich extract and acacia proanthocyanidins-rich extract reduces insulin resistance in 3T3-L1 adipocytes and improves glucose and insulin control in db/db mice. *Nutr Res Pract* 6(5): 405–13.

Trivedi, C.P., N.T. Modi, R.K. Sarin, and S.S. Rao. 1986. Bronchodilator and anti-inflammatory effect of glycosidal fraction of *Acacia farnesiana*. *Indian J Physiol Pharmacol* 30(3): 267–8.

Tseng-Crank, J., S. Sung, Q. Jia, et al. 2010. A medicinal plant extract of *Scutellaria baicalensis* and *Acacia catechu* reduced LPS-stimulated gene expression in immune cells: A comprehensive genomic study using QPCR, ELISA, and microarray. *J Diet Suppl* 7(3): 253–72.

Turiaf, J., J. Tabart, M. Seneze, and D. Vernier. 1959. Allergy to gum arabic in workers in the printing industry. [In French.] *J Fr Med Chir Thorac* 13: 617–35.

Udompornmongkol, P., and B.-H. Chiang. 2015. Curcumin-loaded polymeric nanoparticles for enhanced anti-colorectal cancer applications. *J Biomater Appl* 30(5): 537–46.

Upthegrove, R., and G.M. Khandaker. 2020. Cytokines, oxidative stress and cellular markers of inflammation in schizophrenia. *Curr Top Behav Neurosci* 44: 49–66.

Vanloot, P., N. Dupuy, M. Guiliano, and J. Artaud. 2012. Characterisation and authentication of *A. senegal* and *A. seyal* exudates by infrared spectroscopy and chemometrics. *Food Chem* 135(4): 2554–60.

Wang, L., X. Shen, L. Mi, et al. 2019. Simultaneous determinations of four major bioactive components in *Acacia catechu* (L.f.) Willd and *Scutellaria baicalensis* Georgi extracts by LC-MS/MS: Application to its herb-herb interactions based on pharmacokinetic, tissue distribution and excretion studies in rats. *Phytomedicine* 56: 64–73.

Wapnir, R.A., B. Sherry, C.N. Codipilly, L.O. Goodwin, and I. Vancurova. 2008. Modulation of rat intestinal nuclear factor NF-kappaB by gum arabic. *Dig Dis Sci* 53(1): 80–7.

Wu, J.H., Y.T. Tung, S.C. Chien, et al. 2008. Effect of phytocompounds from the heartwood of *Acacia confusa* on inflammatory mediator production. *J Agric Food Chem* 56(5): 1567–73.

Xiong, J., M.H. Grace, D. Esposito, S. Komarnytsky, F. Wang, and M.A. Lila. 2017. Polyphenols isolated from *Acacia mearnsii* bark with anti-inflammatory and carbolytic enzyme inhibitory activities. *Chin J Nat Med* 15(11): 816–24.

Xiong, J., M.H. Grace, D. Esposito, F. Wang, and M.A. Lila. 2016. Phytochemical characterization and anti-inflammatory properties of *Acacia mearnsii* leaves. *Nat Prod Commun* 11(5): 649–53.

Yimam, M., L. Brownell, M. Pantier, and Q. Jia. 2013. UP446, analgesic and anti-inflammatory botanical composition. *Pharmacognosy Res* 5(3): 139–45.

Yimam, M., P. Jiao, M. Hong, et al. 2018. Repeated dose 28-day oral toxicity study of a botanical composition composed of *Morus alba* and *Acacia catechu* in rats. *Regul Toxicol Pharmacol* 94: 115–23.

Yimam, M., Y.C. Lee, E.J. Hyun, and Q. Jia. 2015a. Reproductive and developmental toxicity of orally administered botanical composition, UP446-Part I: Effects on embryo-fetal development in New Zealand White rabbits and Sprague Dawley rats. *Birth Defects Res B Dev Reprod Toxicol* 104(4): 141–52.

Yimam, M., Y.C. Lee, E.J. Hyun, and Q. Jia. 2015b. Reproductive and developmental toxicity of orally administered botanical composition, UP446-Part II: Effects on prenatal and postnatal development, including maternal function in Sprague-Dawley rats. *Birth Defects Res B Dev Reprod Toxicol* 104(4): 153–65.

Yimam, M., Y.C. Lee, E.J. Hyun, and Q. Jia. 2015c. Reproductive and developmental toxicity of orally administered botanical composition, UP446-Part III: Effects on fertility and early embryonic development to implantation in Sprague Dawley rats. *Birth Defects Res B Dev Reprod Toxicol* 104(4): 166–76.

Yimam, M., Y.C. Lee, P. Jiao, et al. 2016. UP1306, a botanical composition with analgesic and anti-inflammatory effect. *Pharmacognosy Res* 8(3): 186–92.

Yimam, M., Y. Zhao, W. Ma, Q. Jia, S.G. Do, and J.H. Shin. 2010. 90-day oral toxicity study of UP446, a combination of defined extracts of *Scutellaria baicalensis* and *Acacia catechu*, in rats. *Food Chem Toxicol* 48(5): 1202–9.

5 Acacias and Cancer

Cancer has likely been around for a very long time, though not necessarily by that name or according to the modern concept of neoplasia. Usually traditional texts will refer to a disease of swellings or lumps, or, in Latin, a *tumor*.

The idea that cancer is a metaphor for a life gone wrong is also not new. Henry Miller certainly knew it when he named his book *Tropic of Cancer*, because, as he told a confidante, cancer represents a point of desolation, a realization that life had gone wrong, and nothing short of a total change would do (which is what Miller personally attempted in Paris). Various concepts of diseases with swelling and tumors have existed through the ages. And the role of the mind, or even the unity of the mind with the body's physical response to the disease, is also not unknown.

Besides all that, there is an idea that chemicals, even natural chemicals, could act as *carcinogens* and cause cancer. When a likely culprit is named, there is usually a lot of attention to the issue, as the public health is on trial. As the bard sang, ". . . they did a lot of research on it" (Dylan 1989). Curiously, the heartwood of an acacia called *Senegalia catechu* can be extracted with water to make a substance called catechu, among other names. This catechu, with so many excellent anti-inflammatory and anti-bacterial properties, is often chewed along with tobacco and other herbs in a mixture associated with camaraderie and conviviality. At low doses, the substance may be therapeutic, but at higher doses, it may be counter-therapeutic or even pathogenic. The condition in which a drug has such a biphasic effect is called hormesis, or "*hormetic* effect," referring to a process that has been previously described as *potentially toxic at high doses but benign at low doses*. One example, published in peer-reviewed journals, of an hormetic effect is with acacia-derived tannins used in leather tanning (De Nicola et al. 2004; Pagano et al. 2008). Accordingly, notice of the previously described toxicity of acacias leads to further discussion.

In India, such a mixture may be known as "*dohra*" and contains *S. catechu*, areca nut, *i.e.*, the seed of the areca palm, *Areca catechu*, and "edible lime," as well as peppermint leaf (*Mentha piperita*), cardamom seed (*Elettaria cardamomum*), and other "flavorings" along with tobacco (Mishra et al. 2014; Sharma et al. 2018). Unfortunately, the research on the adverse effects of catechu, which uses such ethnographic studies as support, seldom or never breaks down the overall carcinogenic effects of such a mixture to those of the individual components. Most of the time, *Areca catechu* is thought to be the biggest cause of toxicity. However, *S. catechu* has also been suspected because the tannins in *S. catechu* and *A. catechu* may cross some threshold of toxicity and create a harmful synergy (Morton 1992).

In a meta-analysis of 12 studies, *gutka* (the complex of *A. catechu* and *S. catechu*) chewing resulting in severe oral disease in long-time users was documented, with its results including periodontal inflammation (3 studies), oral submucous fibrosis (5 studies), malignant transformation of fibrosis (2 studies), and extension of oral submucous fibrosis into the hypopharynx and esophagus (2 studies) (Javed et al. 2010). A second study of 42 female and 220 male *gutka* chewers revealed that subjects chewing for longer than 10 years had a 2.7% greater chance of developing human papillomavirus (HPV), a genital infection, than chewers for less than 10 years. Further, 78% exhibited pathology, including oral ulcers (25%), rough mucosa (62%), submucosal fibrosis (24%), leukoplakia (20%), and erythroplakia, i.e., idiopathic redness of mucosa (10.6%). The frequency of HPV was highest in patients exhibiting erythroplakia (25%) (Baig et al. 2012).

The anticancer properties of catechu have been noted. One example is the aforementioned proprietary combination of tannin-rich *S. catechu* and flavonoid-rich *Scutellaria baicalensis*. The anti-inflammatory effects of the mixture in terms of the dual inhibition of COX and LOX (cyclooxygenase

DOI: 10.1201/9780429440946-5

and lipoxygenase) are supportive of an anticancer effect through anti-inflammatory mechanisms, as is the inhibition of prostaglandin E2 in COX-2-expressing osteosarcoma cells (Burnett et al. 2007).

Recently, an aqueous extract of *S. catechu* heartwood, standardized to *catechin* and its isomer, *epicatechin*, promoted apoptosis in HT-29 human adenocarcinoma cells, with associated increases in apoptotic cells, reactive oxygen species (ROS), and caspase-9 and caspase-3 activity, and a reduction in mitochondrial membrane potential (MMP). The extract had no effect on normal colon ring function in rats, indicating that the extract's destructive effect was limited to diseased or cancerous tissue (Chiaino et al. 2020).

In vitro, along with three other medicinal plants, *S. catechu* extract inhibited proliferation of cancer cells from the lung (A549), prostate (PC-3), breast (T47D and MCF-7), and colon (HCT-16 and Colo-205). The catechu extract also inhibited leukemia cells (THP-1, HL-60, and K562) and induced, in the K562 type, cell cycle G2/M arrest (Diab et al. 2015).

Also, a 70% methanolic extract of the aqueous *S. catechu* heartwood extract, "pale catechu," known in Hindi as "*katha*," was cytotoxic to estrogen receptor positive MCF-7 breast cancer cells, with an IC(50) of 288.85 ± 25.79 μg/mL. Flow cytometric analysis and morphological assessment were consistent with the induction of apoptosis (Ghate et al. 2014). In addition, the extract was found to be capable of modulating 7,12-dimethylbenz[a]anthracene (DMBA)-induced breast carcinoma in Balb/c mice *in vivo* (Monga et al. 2013).

An *S. catechu* ethanolic seed extract reportedly resulted in cytotoxicity to SCC-25 human tongue squamous cell carcinoma cells, with an IC(50) of 100 μg/mL, upregulating apoptotic markers caspases 8 and 9, cytochrome c, and Bax gene expressions, and with significantly downregulating Bcl-2 gene expression. *Epicatechin*, *rutin*, and *quercetin* were identified in the seed extract (Lakshmi et al. 2017a). A second study of an ethanolic extract of *S. catechu* stem bark indicated cytotoxicity to SCC-25 cells with IC(50) of 52.09 μg/mL (Lakshmi et al. 2017b).

Overall, pure catechu and the other extracts from *S. catechu* seems relatively safe if the dosage is adequately controlled and the extract is taken alone. If combined with other botanical products, though, care must be taken that the patient's total tannin load is not excessive. A yellow catechu dye yields catechin and shows strong specificity for DNA docking sites, suggesting potential for use in designing genetically targeted complex botanical therapies (Hemachandran et al. 2016). A substoichiometric amount of an extract of *S. catechu* enhanced Fenton reaction mediated oxidation in cancer cells but did not damage supercoiled plasmid DNA (as would occur in a standard Fenton reaction). Kar and Chattopadhyaya (2017) note that this finding is especially important when it comes to cancer treatment.

In general, studies of the potential toxicity of *Acacia s.l.* species have so far failed to reveal much cause for concern (Anon 2005). However, one interesting case that has caused considerable discussion was related to two cases of acute toxicity in two males who were college roommates and had together consumed a mixture of *Peganum harmala* seeds and tree bark of an unspecified *Acacia s.l.* species. As a result, a psychotic as well as a somatic acute event occurred, which was thought to be related to dimethyltryptamine. However, the respected author too quickly traversed the pharmacology, assuming that the toxicology was due to the acacia (Liu et al. 2019). Unfortunately the species of *Acacia s.l.* was not specified.

The reported toxicities occurred in Taiwan, and there are only two native species of *Acacia* (though admittedly also some transplants) growing on the island: *Acacia caesia* and *A. confusa*. Of the two, *A. caesia* is reported to contain indoles along with proanthocyanidins, tannins, and lignins (Aruna 2014). *A. confusa*, also known as "Formosa Acacia," is a well-known source of *N,N-dimethyltryptamine* and related tryptamines N-methyltryptamine, N,N-dimethyltryptamine-N-oxide, and N-chloromethyl-N,N-dimethyltryptamine (Buchanan et al. 2007).

Dimethyltryptamine (DMT), as well as other tryptamines, possesses potent and unique anticancer properties, especially influencing the immunological relationship between patient and disease

(Mollica et al. 2012; Tourino et al. 2013). One important trigger for the anticancer activity of DMT is its role as the optimum ligand of the sigma-1 receptor (van Waarde et al. 2015) which plays a key role in the continuation of neoplastic disease (Aydar et al. 2016; Collina et al. 2017; Georgiadis et al. 2017; Huang et al. 2014; Soriani et al. 2017). In addition to the production of DMT by cells of a melanoma line, such cells also secreted peroxidases, which oxidized additional DMT added to the medium (Gomes et al. 2014). This is puzzling, but apparently DMT is part of this cancer cell's metabolism, and further, there is a rational basis for the emerging use of DMT and/or related tryptamines in anticancer therapy (Szabo 2015; van Waarde et al. 2010). There is no doubt that *A. confusa* has anticancer effects, mostly from the proteins in its seeds but also from its DMT on sigma receptors.

Acaconin, an antifungal protein purified from *A. confusa* seeds, inhibited proliferation of MCF-7 breast cancer cells (IC(50) 128 ± 9) but did not affect hepatoma HepG2 cells (Lam and Ng 2010a, b). *Concanavalin A* and a trypsin inhibitor isolated from *A. confusa* were conjugated to N-succinimidyl-3-(2-pyridyldithio)propionate, and the conjugate, when injected once into mice bearing sarcoma 180, exerted a "remarkable effect of increasing the survival" (Lin and Lin 1985). The discovery of an abrin-b chain helped establish a mechanism for intracellular transport of the active component (Lin et al. 1989).

Avicins are triterpene saponins, originally extracted from the Australian desert acacia *Acacia victoriae*, with exceptional anti-cancer cell selectivity even at very low doses, e.g., 0.2 µg/mL (Joshi et al. 2002). Avicins were early identified by Valsala Haridas and team at the M.D. Anderson Cancer Center in Houston, Texas, as pro-apoptotic agents against cancer cells with a mechanism of perturbation to the cancer cell's mitochondrial membranes (Haridas et al. 2001b), most specifically against mitochondrial outer membranes (Haridas et al. 2007). The anti-inflammatory root of the anti-cancer effect is exemplified by findings that revealed avicins' biochemical finesse against the key inflammatory mediator, NFkappaB, inhibiting both its nuclear localization and its ability to bind DNA (Haridas et al. 2001a). Avicins also prevented chemical carcinogen-induced skin carcinoma in mice (Hanausek et al. 2001). In-depth work with one of the most promising compounds showed *avicin D* was capable of inducing autophagy (self-eating) by activation of an AMP-activated protein kinase (Xu et al. 2007), stimulating apoptosis by recruitment of Fas and downstream signaling compounds into lipid rafts (Xu et al. 2009), and dephosphorylating Stat-3 by regulating activities of both kinase and phosphatase (Haridas et al. 2009).

One of the avicins, *avicin D*, has been the object of considerable attention. Its anticancer properties are linked to its anti-inflammatory properties, which it shares with the prototypical steroid glucocorticoid, for example, dexamethasone. Avicin D blocks and binds to the glucocorticoid receptor, but its cancer-selective pro-apoptotic function is spared even when its glucocorticoid function is blocked (Haridas et al. 2011) (see Figure 5.1).

Elsewhere, avicin D selectively induced apoptosis in cutaneous T-cell lymphoma cells while downregulating survivin, phosphorylated signal transducer and activator of transcription 3 protein (p-STAT-3), the latter from the STAT-3 gene, and B-cell lymphoma protein bcl-2 (Zhang et al. 2008). The pro-apoptotic effect of avicins may involve internal cell machinery using the small protein *ubiquitin* to suppress heat shock protein 70 and X-linked inhibitors of apoptosis proteins (Gaikwad et al. 2005). Mujoo et al. (2001) found that *Avicin D* and *Avicin G*, both triterpenoid saponins from *A. victoriae*, caused the human MDA-MB-453 breast cancer cell line to stop in the G1 phase of the cell cycle and the Jurkat (T-cell leukemia) and MDA-MB-435 breast cancer cell lines to die.

In general, the mechanism for avicins' selective cytotoxicity is increasing the solubility of the cancer cells' mitochondrial membranes (Lemeshko et al. 2006), causing diminution of the tumor cells' utilization of energy (Haridas et al. 2007). Haridas et al. (2004) found that avicins activate an innate stress response through redox regulation of a gene battery. Croce (2001) found that avicins may have a double effect on cancer biology by both promoting apoptosis and changing the way cells' genes lead them to become cancerous.

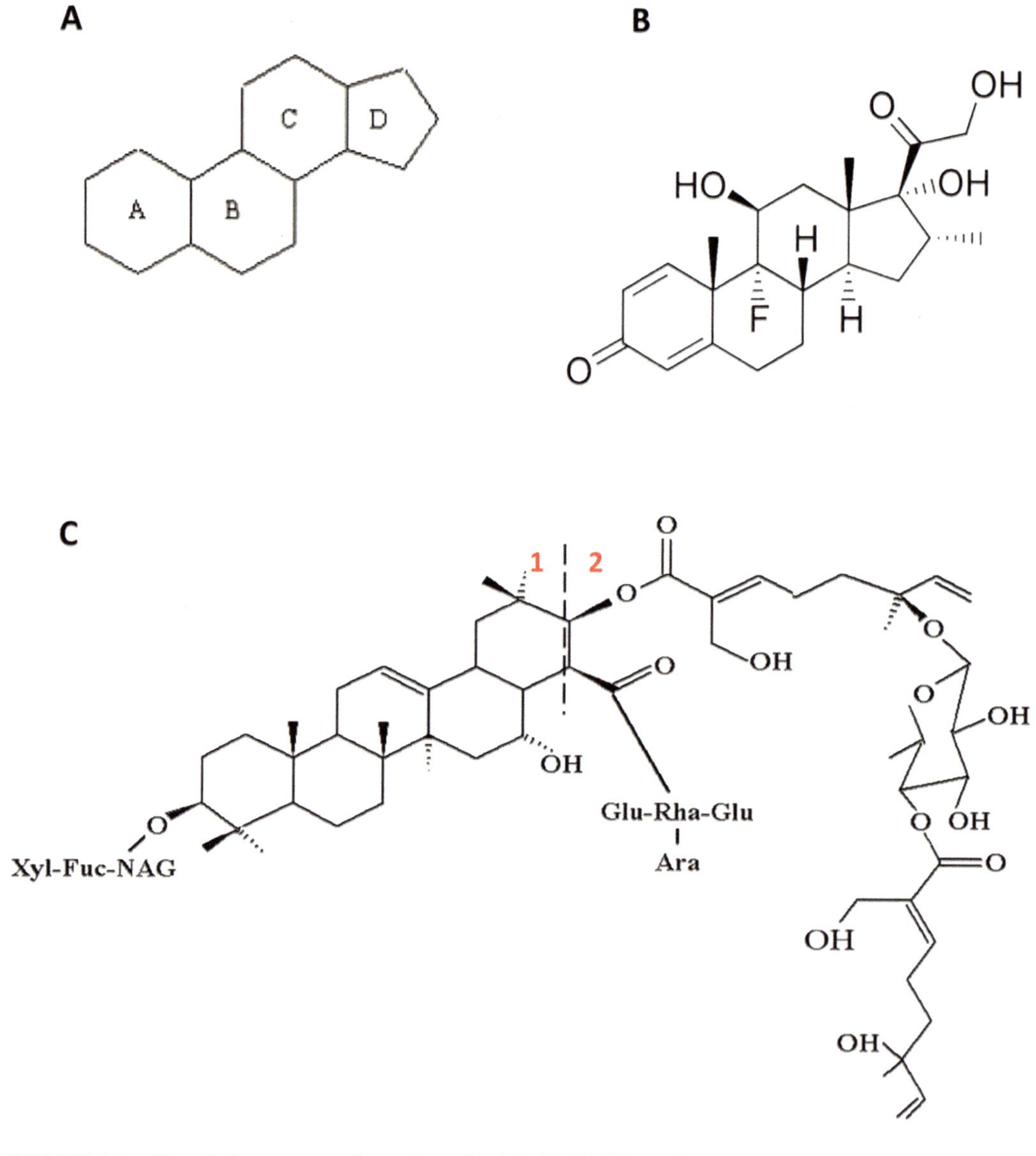

FIGURE 5.1 Chemical structures of steroids and avicin D. (A) The basic ring structure of a steroid molecule. (B) Chemical structure of dexamethasone, a prototypical steroid. (C) Chemical structure of avicin D. Part 1 of the molecule has the core 5-ring structure, which resembles the core structure of a steroid molecule, and Part 2 has a side chain containing two units of acyclic monoterpenes, connected by a quinovose sugar. (Haridas et al. 2011, CC BY, https://doi.org/10.1371/journal.pone.0028037.g001.)

Although the case of avicins is exceptional for the specificity of the cytotoxicity, and its mechanism is being increasingly ascertained, it also connects with the wider concept of utilizing saponins in general, with the prime example of avicins, for complementary or stand-alone anti-cancer therapy (Bachran et al. 2008; Man et al. 2010; Yadav et al. 2010). But according to Gutterman et al. (2005), even though the treatment is generally safe, there may be safety concerns that need more attention.

Vachellia nilotica, the Israeli babul, has also been looked to for potential anticancer activities. According to one computer modeling, three activated compounds within *V. nilotica*, namely kaempferol, ellagic acid, and quercetin, were predicted to have the best activity together and had the potential to be used in that way as viable anticancer formulations (Al-Nour et al. 2019). Curiously, those three polyphenols (two flavonoids and one tannin) also emerged as signal compounds in previous anticancer investigations of pomegranate fruit, *Punica granatum* (Lansky et al. 2005; van Elswijk et al. 2004). However, a later analytical study of *V. nilotica* seed pods found the aqueous, ethyl acetate, and N-butanol fractions to contain tannins, saponins, flavonoids, and carbohydrates but no anthraquinones, alkaloids, terpenes, or steroids. Minerals present included iron, potassium, manganese, zinc, calcium, phosphorus, magnesium, sodium, cadmium, and copper (Auwal et al. 2014).

Barapatre et al. (2016) used a series of solvent extractions to get lignin fractions from *V. nilotica* heartwood. These fractions were toxic to MCF-7 breast cancer cells (IC(50) ≥ 100 μg/mL), but not to normal primary human hepatic stellate cells.

V. nilotica was one of the four most potent plants out of 900 tested for inhibiting lactate dehydrogenase A, an enzyme associated with malignant disease and an appealing oncologic target but with few known inhibitors (Deiab et al. 2013). Ethanol was the most effective solvent for extracting anticancer principles from *V. nilotica* leaves, which were cytotoxic against Vero (green monkey kidney epithelial) and Hela cells (human cervical cancer), with IC(50)s of 53.6 and 28.9 μg/mL, respectively (Kalaivani et al. 2011).

An acetone extract of *V. nilotica* bark was more effective than its chloroform extract and more effective than either the chloroform or acetone extracts of *Acacia auriculiformis* bark in reducing the occurrence of breast tumors in response to the carcinogen 7,12-dimethylbenz[a]anthracene (DMBA) in an *ex vivo* model of mouse mammary organ culture (Kaur et al. 2002). *In vivo*, Swiss male albino mice provided oral gavage of an aqueous extract of *V. nilotica* leaves showed strong prevention of the development of skin papillomagenesis induced by DMBA, which was superior to that associated with similar extracts obtained from *V. nilotica* flowers that were superior to those from *V. nilotica* gum (Meena et al. 2006). *V. nilotica* leaf extract, as well as one of its components, *ethyl gallate*, was effective in halting the growth of oral squamous cell carcinoma in a mouse model without toxicity to the animal (Mohan et al. 2014, 2017). *Pyrogallol*, another *gallic acid* derivative, was extracted from *V. nilotica* and was more effective than the extract itself—which was also effective—in inducing a "dead state" of advanced apoptosis and DNA fragmentation in HT-29 human colon cancer cells (Revathi et al. 2018). A *V. nilotica* extract, injected *i.p.* into animals on selected days prior to chemical induction of Dalton's ascitic lymphoma (DAL), resulted in a more marked and significant reduction of tumor mass and associated biochemical factors than controls (Sakthivel et al. 2012).

Silver nanoparticles synthesized with *V. nilotica* gum as a reducing and stabilizing agent were cytotoxic to LoVo cells (epithelial cells originating from a metastatic colon cancer) with an IC(50) of 35–57 μg/mL (Mohammed et al. 2018). Silver-doped titanium oxide nanoparticles reduced and stabilized with *V. nilotica* gum inhibited MCF-7 breast cancer cell proliferation *in vitro* (Rao et al. 2019).

HT-29 cells are more sensitive to the *V. nilotica* compound pyrogallol, with an IC(50) of 35 μg/mL compared to ACE. Pyrogallol-treated HT-29 cells reached a "dead state," i.e., a late apoptotic state with severe nuclear fragmentation. Pyrogallol elicits dose-dependent DNA fragmentation in HT-29 cells. It also induced apoptosis by simultaneous downregulation of Bcl-2 and upregulation of BAX and cytochrome c, arrested HT-29 cells in the S and G2/M phases of their cell cycle, and exhibited marked antimetastatic potential by dose-dependently inhibiting the migration of HT-29 cells (Revathi et al. 2018).

An extract of *V. nilotica* bark, given orally once per day for 10 weeks, strongly and significantly prevented liver damage in rats with hepatocellular carcinoma that had been induced by N-nitrosodiethylamine injection. Chromatographic analysis (HPLC) of the extract revealed *quercetin* and *gallic*, *protocatechuic*, *caffeic*, and *ellagic acids* (Singh et al. 2009). Another study discovered γ-sitosterol in a *V. nilotica* leaf extract and demonstrated that the extract and γ-sitosterol isolated from the extract, had anticancer activity against MCF-7 breast cancer cells, A-549 lung cancer cells, and promoted apoptosis. Cell cycle arrest occurred at phase G2/M and c-Myc expression was downregulated (Sundarraj et al. 2012). When consumed as a medicinal food, the seeds of *V. nilotica* may also provide anticancer activity (Vadivel and Biesalski 2012). *V. nilotica* was one of 13 higher plant species out of 1,600 to suppress histone deacetylases, enzymes related to the progression of disease in advanced malignancy (Mazzio and Soliman 2017).

Senegalia senegal, in the form of "gum arabic," is used in many formulations as a stabilizer or vehicle, and it may also contribute to anticancer effects when used with other plants. For example, microspheres containing gum arabic and curcumin are able to solve some of the inherent difficulty of delivering curcumin to tissues, in this case for the treatment of triple negative breast cancer (Pal et al. 2019, 2020). Still, some evidence suggests that gum arabic may also have anticancer effects of its own, which can be safely, inexpensively, and confidently combined with other therapies. In one study, chemical carcinogenesis resulted in cancerous tumors of the colon in mice. When some of the mice received 10% by weight gum arabic in their drinking water, the occurrence of tumors dropped by 70% (Nasir et al. 2010). Gum arabic also inhibited the pro-angiogenic enzymes angiogenin 1, 3, and 4, thereby inhibiting colon angiogenesis, which is linked to the progression of colon cancer (Nasir 2013). A complex of gum arabic and gold exerted cytotoxic as well as anti-angiogenic properties (Yan et al. 2010). Gum arabic protected the urinary bladder from toxicity secondary to the oncologic cytotoxic drug cyclophosphamide (Al-Yahya et al. 2009). Gum arabic, when combined with *Lactobacillus plantarum*, acted as a prebiotic and reduced *in vivo* TNF-α (tumor necrosis factor), a pro-inflammatory marker associated with cancer progression in general (Chundakkattumalayil et al. 2019).

Vachellia hydaspica (aka *V. eburnea*) is known as a medicinal source in India and has been studied for its anticancer properties. Most of the investigative work in this highly specialized field was led by Dr. Tayyad Afsar and her associates at King Saud University in Riyadh. For example, *V. hydaspica* extracts suppressed the proliferation of both androgen receptor independent PC-3 prostate cancer cells and ER—MDA-MB-231 breast cancer cells. Active compounds identified in the extract through bioactivity-guided fractionation included *7-O-galloyl catechin* (GC), *catechin* (C), *methyl gallate* (MG), and *catechin-3-O-gallate* (CG). All four compounds inhibited the PC-3 cells, but only the latter two (MG and CG) inhibited the MDA-MB-231 (Afsar et al. 2016). A parallel study stressed the antioxidant properties of the extract and identified *gallic acid* and *myricetin* (Afsar et al. 2016). Afsar and co-workers further found the extract, due to its potent antioxidant activities, to prevent or ameliorate testicular damage in rats caused by the cancer chemotherapeutic cisplatin (Afsar et al. 2017). The most potent antioxidant and anticancer compounds were obtained in the highest yield from the ethyl acetate fractions of the *V. hydaspica* extracts (Afsar et al. 2018).

It is important to remember that acacias often bear edible seeds, and both the seed pods and the seeds themselves may contain medicinal principles. Searching for leukemia-specific leukocyte-agglutinating lectins among the seeds of a panel of plant seeds, "saline extracts" of the seeds of all ten plants in the series agglutinated types of leukemia cells but not normal leukocytes or lymphocytic disease cells were found. Among the group were the seeds of *Senegalia lenticularis* and *S. catechu*. (The other eight species were *Cassia marginata, Cicer arietinum, Crotalaria juncea, Ficus racemosa, Gossypium indicum, Melia composita, Millettia ovalifolia*, and *Peltophorum ferrugineum*.) In every case, clumping could be prevented by adding simple sugars (Agrawal and Agarwal 1990).

A comparative study from Saudi Arabia tested the cytotoxicity of ethanolic aerial part extracts from four different acacias grown in the Kingdom, *Acacia salicina, Senegalia hamulosa, Senegalia*

FIGURE 5.2 *Senegalia lenticularis*, *chah*, is a small, deciduous, and moderately thorny tree. It is cultivated for shade and for timber, which is very hard and of excellent quality. *Chah* occurs in India, Pakistan, and Nepal. One tree produces 1 kg/year of gum (Gamble 1902). Illustration: *S. lenticularis* leaf, drawing. (In Dietrich Brandis, *Indian Trees: An Account of Trees, Shrubs, Woody Climbers, Bamboos, and Palms Indigenous or Commonly Cultivated in the British Indian Empire* [London: A. Constable & Co., 1906], p. 297, public domain, via Flickr, Internet Archive Book Images, www.flickr.com/photos/internetarchivebookimages/20618527638.)

laeta, and *Vachellia tortilis*, individually, against human cancer cells of the liver (Hep G2), kidney (HEK 203), and breast (MCF-7, MDA-MB 231). The results are summarized in Table 5.1. *Hint:* the lower values reflect greater cytotoxicity (Alajmi et al. 2017).

TABLE 5.1

Estimated Cytotoxicity of Four Acacias against Human Cancer Cells (as IC(50) in μg/mL, to nearest integer)

Acacia species	HepG2 (Liver)	HEK-293 (Kidney)	MCF-7 (ER+ Breast)	MDA-MB-231 (ER- Breast)
Acacia salicina	90	75	98	104
Senegalia hamulosa	39	93	55	59
Senegalia laeta	46	54	57	60
Vachellia tortilis	42	49	66	52
5-fluorouracil (positive control)	3	2	4	4

Source: After Alajmi et al. (2017).

(a) (b) (c)

FIGURE 5.3 The bark of *Acacia salicina*, "Sally wattle" of Australia, was used in making a fish poison by indigenous peoples. The wood is used for fine furniture and for burning. The seeds are edible. (a) *A. salicina*, tree trunk and bark. (9/22/2019, Woodstock-Cleveland-Ross, Reid River, Queensland, Australia: by Elawrey, CC BY 4.0, via iNaturalist, www.inaturalist.org/photos/52725822.) (b) *A. salicina*, mature seed pods, seeds, and foliage. (9/22/2019, Woodstock-Cleveland-Ross, Reid River, Queensland, Australia: by Elawrey, CC BY 4.0, via iNaturalist, www.inaturalist.org/photos/52725801.) (c) *A. salicina* twig with buds, flowers, and leaves. (2/7/2020, New Queen Road, Queenton, Queensland, Australia: by Elawrey, CC BY 4.0, via iNaturalist, www.inaturalist.org/photos/61049660.)

Methanolic and ethyl acetate extracts of *A. salicina* significantly suppressed the proliferation *in vitro* of cells of K562 human chronic myelogenous leukemia and L1210 murine leukemia cell lines (Chatti et al. 2009).

A novel oleanane-type triterpenoid saponin, *polyacanthoside A*, isolated from *Senegalia polyacantha* leaves and stem bark, showed cytotoxic activity against CCRF-CEM leukemia cells at IC(50) 8.90 μM and HepG2 hepatocarcinoma cells IC(50) 35.21 μM (Fotso et al. 2019).

Leaves of *Acacia saligna*, also known as *Acacia cyanophylla* owing to the blue tint of its leaves, yielded a novel spirostane saponin *(25S)-5β-spirostan-3β-yl-3-O-β-D-xylopyranosyl(1 → 3)-O-β-D-xylopyranosyl(1 → 4)-β-D-galactopyranoside* and biflavonoid glycoside *myricetin-3-O-rhamnoside (C7-O-C7) myricetin-3-O-rhamnoside* along with known compounds *erythrodiol*,

FIGURE 5.4 *Senegalia laeta*, "gay acacia" or daga, is used in the making of dye and a *faux* gum arabic, as well as fencing for *bomas* in Africa. The tree grows to a height of 4–10 m, and it has distinctive bark with an iridescent pink tint. (a) *S. laeta* tree. (9/7/2006, Kongoussi, Burkina Faso: by Dr. Ron Frumkin, with permission, via iNaturalist, www.inaturalist.org/photos/112022293.) (b) *S. laeta*, flower and seed pod. (9/7/2006, Kongoussi, Burkina Faso: by Dr. Ron Frumkin, with permission.) (c) *S. laeta*, flower buds, bark, and foliage. (9/7/2006, Kongoussi, Burkina Faso: by Dr. Ron Frumkin, with permission.) (d) *S. laeta*, seed pods. (9/7/2006, Kongoussi, Burkina Faso: by Dr. Ron Frumkin, with permission, via iNaturalist, www.inaturalist.org/photos/112022304.)

3β-O-trans-p-coumaroyl-erythrodiol, *quercetin-3-O-α-L-rhamnoside*, and *myricetin-3-O-α-L-rhamnoside*. The compounds and the extract were active against HEPG2 liver cancer cells *in vitro* (Gedara and Galala 2014).

Bioactive fractionation was used to isolate both new and known compounds from 70% acetone extracts of *Acacia burkittii* and *A. acuminata*. The compounds isolated included two flavan-3,4-diols, *melacacidin* and *isomelacacidin*, and three flavonoids, *3,7,8,3',4'-pentahydroxyflavone*, *7,8,3',4'-tetrahydroxyflavanone*, and *3,7,8,3',4'-pentahydroxyflavanone*. While non-toxic to "normal" CV-1 kidney cells from green monkeys, the compounds were all toxic to cancerous L1210 mouse lymphocytic leukemia cells. Compared to a panel of ten commercially available phenolic compounds, including catechin and epicatechin, the flavan-3,4-diols killed cancer cells more effectively than any of the commercial phenolics (Grace et al. 2009).

(a) (b) (c)

FIGURE 5.5 Native to Africa, India, the Indian Ocean, and Asia, *Senegalia polyacantha* [= "multi-thorned"], white thorn, is a substantial tree, growing to 25 m. It is used as lumber, but also as medicine for snakebite and venereal diseases, and for washing the skin of children during nighttime agitations. (a) *S. polyacantha* tree. (2/26/2017, Muthanallur Lake, Bommasandra, Karnataka, India: by Ajit Ampalakkad, CC0 1.0, via iNaturalist, www.inaturalist.org/photos/157729206.) (b) *S. polyacantha*, leaves and a mature seed pod. (2/26/2017, Muthanallur Lake, Bommasandra, Karnataka, India: by Ajit Ampalakkad, CC0 1.0, via iNaturalist, www.inaturalist.org/photos/157729287.) (c) *S. polyacantha*, trunk with bark and thorns. (2/26/2017, Muthanallur Lake, Bommasandra, Karnataka, India: by Ajit Ampalakkad, CC0 1.0, via iNaturalist, www.inaturalist.org/photos/157729452.)

(a) (b)

FIGURE 5.6 *Acacia burkittii*, Burkitt's wattle, is a spreading or erect shrub of 1–4 m in height. It has finely fissured bark and terete, glabrous branchlets. It is known for containing alkaloids, including DMT and N-methyltryptamine, in its bark and aerial parts (White 1957). Formerly, it was considered a subspecies of *A. acuminata*, but is now classified as a separate species. (a) *A. burkittii* branch with phyllodes and yellow flowers. (8/31/2020, Carriewerloo SA 5715, Australia: by Kym Nicolson, CC BY 4.0, via iNaturalist, www.inaturalist.org/photos/93673459.) (b) *A. Burkittii* F.v.M., botanical drawing. (By M. Flockton, in J. H. Maiden, *The Forest Flora of New South Wales*, Vol. 6, Table 224 [Sydney: W. A. Gullick, 1913], via www.plantillustrations.org, public domain, www.plantillustrations.org/illustration.php?id_illustration=63225.)

(a) (b)

FIGURE 5.7 *Acacia acuminata*, "jam wattle," which may grow as high as 7 m, is widely distributed throughout southwestern Australia. The durable and attractive wood, when freshly cut, is said to smell like raspberry jam, hence the tree's nickname. (a) *A. acuminata*, phyllodes and yellow flowers. (7/12/2020, Tampu, Western Australia, Australia: by Loxley Fedec, CC BY-ND 4.0, via iNaturalist, www.inaturalist.org/photos/85942636.) (b) *A. acuminata*, a blossoming shrub. (7/12/2020, Tampu, Western Australia, Australia: by Loxley Fedec, CC BY-ND 4.0, via iNaturalist, www.inaturalist.org/photos/85942627.)

(a) (b)

FIGURE 5.8 *Acacia kempeana*, "Wanderrie wattle," grows as a spreading tree or shrub, usually 1–6 m in height, though trees as high as 10 m have been recorded. The tree has also been called "witchetty bush" for the grubs that feed on the roots which are also eaten by Australian Aboriginal peoples. (a) *A. kempeana* shoots with flowers and leaves. (1/23/2015, Watarrka National Park, Northern Territory, Australia: by Kenneth Bader, with permission, via iNaturalist, www.inaturalist.org/photos/1586055.) (b) *A. kempeana* F.Muell, flowers, leaves, seed pods, and seeds; botanical drawings. (In F. J. H. von Mueller, *Iconography of Australian species of* Acacia *and cognate genera*, Table 99 [Melbourne: J. Ferres, Govt. Printer, 1887–88], via www.plantillustrations.org, public domain, www.plantillustrations.org/illustration.php?id_illustration=210483.)

Ethanolic extracts of *Acacia kempeana* and *Acacia tetragonophylla* showed potent and specific activity against HeLa cells, thus suggesting their possible utility in the treatment of cervical cancer (Gulati et al. 2015).

Robinetinidol-(4β-->8)-epigallocatechin 3-O-gallate, a galloyl dimer prorobinetinidin purified from *Acacia mearnsii*, protects neuroblastoma cells from oxidative damage associated with the chemical toxin acrolein by suppressing ROS production. Thus, it is necessary to consider such factors when designing herbal cytotoxic regimens for treating patients with cancer in order to harmonize and not conflict with the anticancer purpose of the therapy (Huang et al. 2010).

Four new diterpenes, *acasiane A*, *acasiane B*, *farnesirane A*, and *farnesirane B*, along with three known diterpenes, two triterpenes, and eight flavonoids, were isolated from the roots of *Acacia farnesiana* and tested for possible cytotoxic activity against cells of selected human cancer cell lines, namely HepG2, Hep3B, MDA-MB-231, MCF-7, A549, and Ca9–22 (human gingiva carcinoma). Unfortunately, complex extracts were not employed, and the investigation relied solely on purified compounds. Further exploration employing complex *A. farnesiana* root extracts was indicated (Lin et al. 2009). Since then, *diosmetin*, derived from *A. farnesiana* leaves (as well as from leaves of *Olea europaea*), has been shown to be cytotoxic and to induce apoptosis in ACHN renal cancer cells via downregulated phosphorylation of the phosphoinositide 3-kinase and protein B kinase (PI3K/AKT), in other words, by reducing AKT phosphorylation through p53 upregulation (Qiu et al. 2020).

Acacia pennivenia, the Socotran acacia, was tested along with other Socotran higher plant species as possible sources of inhibitors of selected human malignancies. The methanolic extract of *A. pennivenia* was listed in the results of the study as simply >50 μg/mL against cancer cell lines 5637 (urinary bladder), MCF-7 (breast), and A-427 (lung) (Mothana et al. 2009).

An extract of the Mexican species *Vachellia pennatula* was cytotoxic to cells of KB, HCT-15 COLADCAR and UISO-SQC-1 cell cultures (Popoca et al. 1998).

(a) (b)

FIGURE 5.9 A species of Mexico, the beautiful tree *Vachellia pennatula* may grow to heights of 8 m. (a) *V. pennatula* branch with leaves and seed pods. Note also the small thorns. (10/31/2021, Camino a Lagunillas, Tapalpa, Jalisco, Mexico: by Arnoldo Uribe Zamora, CC BY-ND 4.0, via iNaturalist, www.inaturalist.org/photos/167448993.) (b) *V. pennatula* tree, habitat. (7/10/2021, Oaxaca, Mexico: by Elliot Greiner, CC BY 4.0, via iNaturalist, www.inaturalist.org/photos/142472547.)

(a) (b)

FIGURE 5.10 *Senegalia macrostachya*, *bihongaga* (in the Balanta language), is a well-known tree in the regions of Senegal where it grows. Its seeds and pods are used for food and its bark as medicine for pain, diarrhea, venereal disease, and stomach problems, including emesis, and as an arrow poison. (a) *S. macrostachya* branches with seed pods. (CC BY 4.0, via Tsammalex: A lexical database on plants and animals [http://tsammalex.clld.org], https://cdstar.shh.mpg.de/bitstreams/EAEA0-86A8-1FB9-A829-0/full.JPG.) (b) *S. macrostachya*, tree trunk and bark. (CC BY 4.0, via Tsammalex: A lexical database on plants and animals [http://tsammalex.clld.org], https://cdstar.shh.mpg.de/bitstreams/EAEA0-A5DD-77FE-31E3-0/full.JPG.)

From *Acacia mellifera* bark were derived novel pentacyclic terpenoids *(20R)-3-oxolupan-30-al*, *(20S)-3-oxolupan-30-al*, and *(20R)-28-hydroxylupen-30-al-3-one*, as well as the known *(20S)-3beta-hydroxylupan-30-al*. All but *(20R)-28-hydroxylupen-30-al-3-one* were inactive. This compound, however, inhibited human non-small-cell lung cancer cell line cells with an IC(50) of 39.5 ± 1.2 μM. Other compounds observed in the extracts included the known metabolites *30-hydroxylup-20-(29)-en-3-one*, *30-hydroxylup-20-(29)-en-3beta-ol*, *atranorin*, *methyl 2,4-dihydroxy-3,6-dimethyl benzoate*, and *sitosterol-3beta-O-glucoside*, as well as *linoleic acid*, which had apparently been found in *A. mellifera* for the first time (Mutai et al. 2007).

Two terpenoids and a flavonoid glycoside were isolated from *Senegalia pennata* and proven to inhibit the Hedgehog/glioma-associated oncogene, as well as exhibiting selective cytotoxicity against human pancreatic (PANC1) and androgen-negative prostate (DU145) cancer cells *in vitro* (Rifai et al. 2010).

Two monoterpenoid glycosides, *cyclopside 1* and *cyclopside 2*, were isolated from *Acacia cyclops*. Cyclopside 1 was more potently cytotoxic, causing a 91% suppression of growth of the MCF-7 breast cancer cells at 50 μg/mL (Jelassi et al. 2014).

Working with melanoma and lymphoma cell cultures, researchers showed *Senegalia ferruginea* extracts to inhibit cell survival and prevent angiogenesis and tumor growth while suppressing inflammatory mediators TNF-α, iNOS, COX-2, IL-1β, IL-6, IFN-γ, IL-2, and GM-CSF *in vivo* (Sakthivel and Guruvayoorappan 2013, 2018).

An extract of *Acacia macrostachya* exerted a potent 95% antiproliferative effect on KB cancer cells, with IC(50) of 4.30 ± 0.26 μg/mL, while not impairing normal Vero and MCR5 cells (Sawadogo et al. 2012).

Bioactive fractionation of a methanolic extract of *Senegalia tenuifolia* led to the discovery of three new and three known saponins, part of which exhibited cytotoxicity against selected mammalian cell lines (Seo et al. 2002).

Proanthocyanidins extracted from the bark of *Acacia mearnsii* had varying effects on cancer cells, depending on the solvent used in the extractions. For example, the ethyl acetate fractions effectively inhibited proliferation of ER- human breast cancer cells (MDA-MB-231) and human liver cancer cells (BEL-7402) and had a weak effect on human cervical cancer cells (HeLa) but

(a)

(b)

(c)

FIGURE 5.11 *Senegalia tenuifolia*, *tocino*, is a climbing shrub that can reach heights of 8 m with diameters of 15 cm. It grows in Mexico, Central America, and the northern and central parts of South America. (a) *Acacia paniculata* (= *S. tenuifolia* (L.) Britton & Rose), botanical drawing. (In C. F. P. von Martius, A. G. Eichler, and I. Urban, *Flora Brasiliensis*, Vol. 15: *Leguminosae II et III, Swartzieae, Caesalpinieae, Mimoseae* [Monachii et Lipsiae: R. Oldenbourg, 1870–1876], public domain, via www.plantillustrations.org, www.plantillustrations.org/illustration.php?id_illustration=14297.) (b) *S. tenuifolia*, immature seed pod. (10/24/2002, Barro Colorado Island, Panama Canal, Panama: by Steven Paton, public domain, Smithsonian Tropical Research Institute, https://biogeodb.stri.si.edu/bioinformatics/dfmfiles/files/c/7212/7212.jpg.) (c) *S. tenuifolia* var. *tenuifolia* branch with leaves and yellow flowers. (12/2/2021, Imaculada, Paraíba, Brazil: by Willa Rodrigues, with permission, via iNaturalist, www.inaturalist.org/photos/171108349.)

(a) (b)

FIGURE 5.12 *Vachellia schaffneri*, twisted acacia or Schaffner's acacia, is native to Mexico and Texas. The wood is used for fuel and fences. Goats and sheep graze on the foliage. (a) *V. schaffneri* trees, habitat. (11/5/2015, Tultitlán, MX, Mexico: Fernando Cecor, CC BY 4.0, via iNaturalist, www.inaturalist.org/photos/2615305.) (b) *V. schaffneri*, branches with seed pods and leaves. (5/8/2021, Dolores Hidalgo Cuna de la Independencia Nacional, Guanajuato, México: by Juan Carlos Fonseca Mata, CC BY 4.0, via iNaturalist, www.inaturalist.org/photos/127549913.)

(a) (b)

FIGURE 5.13 *Vachellia seyal*, red acacia, is also known as the "shittah tree" because of its putative identity as one of the acacias that, according to the Torah, supplied the wood for the Ark of the Covenant. Its gum is called, based on the Arabic name, gum *talha*. The bark was used for treating skin diseases, possibly including leprosy. The gum is an aphrodisiac, and the incense made from burning the wood is said to be good for those suffering from rheumatism. (a) *V. seyal* tree. (5/13/2015: by TreeWorld Wholesale [www.treeworldwholesale.com], CC BY 2.0, via Flickr, www.flickr.com/photos/treeworld/34844845196.) (b) *V. seyal* branches: leaves, thorns, and the characteristic reddish-brown bark. (9/29/2016: by TreeWorld Wholesale [www.treeworldwholesale.com], CC BY 2.0, via Flickr, www.flickr.com/photos/treeworld/34844846736/.)

no effect on human lung cancer cells (A549). Shen et al. (2010) found that the aqueous extract of the same bark had a mild effect on MDA-MB-231, HeLa, and A549. However, it had no effect on BEL-7402.

Three *seco-oxacassanes* (1–3) were isolated from *Vachellia schaffneri*. Seco-oxacassane 2 was extremely active against colon (HT-29), lung (A549), and melanoma (UACC-62) cancer cells in culture, with IC(50) values ranging from 0.12 to 0.92 g/mL (Torres-Valencia et al. 2015).

A hydroethanolic extract of *Vachellia seyal* induced apoptosis in MDA-MB-231 estrogen receptor-negative breast cancer cells but failed to do so in MCF-7 estrogen receptor positive breast cancer cells. In the MDA-MB-231 cells, the extract induced cell cycle arrest at G2/M and inhibited invasion (Zingue et al. 2018a). The same hydroalcoholic extract of *V. seyal* also significantly decreased the incidence and volume of mammary tumors induced in mice by the carcinogen DMBA (Zingue et al. 2018b).

An ethanolic extract of the seed pods of *Acacia ligulata* yielded two new echinocystic acid triterpenoid saponins, *ligulataside a* and *ligulataside b*. Both compounds were weakly cytotoxic against SK-MEL28 human melanoma cells (Jaeger et al. 2017, 2018).

(a)

(b)

FIGURE 5.14 *Acacia ligulata*, also known as "umbrella bush" or "sandhill wattle," is a spreading shrub with a width of about 3 m and a height of 2–4 m. One of the most common *Acacia sensu stricto* species in Australia, it is located mainly in the central and southern regions. Its roots host the witchetty grub and the larvae of the *Nacaduba biocellata* butterfly. (a) *A. ligulata* bush. (1/10/2021, Colignan VIC 3494, Australia: by Dylan Butcher, CC BY 4.0, via iNaturalist, www.inaturalist.org/photos/109806173.) (b) *A. ligulata* branches with seed pods. (1/10/2021, Colignan VIC 3494, Australia: by Dylan Butcher, CC BY 4.0, via iNaturalist, www.inaturalist.org/photos/109806202.)

(a)

(b)

FIGURE 5.15 *Nacaduba biocellata*, a two-spotted line-blue butterfly. (a) *N. biocellata* on *Styphelia* flowers. (6/18/2021: by Jean and Fred Hort, CC BY 2.0, via Flickr, www.flickr.com/photos/jean_hort/51267773464.) (b) *N. biocellata*. (11/28/2010, Witera SA 5671, Australia: by Rolf Lawrenz, CC BY 4.0, via iNaturalist, www.inaturalist.org/photos/24869432.)

REFERENCES

Afsar, T., S. Razak, M.R. Khan, et al. 2016. Evaluation of antioxidant, anti-hemolytic and anticancer activity of various solvent extracts of *Acacia hydaspica* R. Parker aerial parts. *BMC Complement Altern Med* 16: 258.

Afsar, T., S. Razak, M.R. Khan, and A. Almajwal. 2017. *Acacia hydaspica* ethyl acetate extract protects against cisplatin-induced DNA damage, oxidative stress and testicular injuries in adult male rats. *BMC Cancer* 17(1): 883.

Afsar, T., S. Razak, M. Shabbir, and M.R. Khan. 2018. Antioxidant activity of polyphenolic compounds isolated from ethyl-acetate fraction of *Acacia hydaspica* R. Parker. *Chem Cent J* 12(1): 5.
Agrawal, S., and S.S. Agarwal. 1990. Preliminary observations on leukaemia specific agglutinins from seeds. *Indian J Med Res* 92: 38–42.
Al-Nour, M.Y., M.M. Ibrahim, and T. Elsaman. 2019. Ellagic acid, kaempferol, and quercetin from *Acacia nilotica*: Promising combined drug with multiple mechanisms of action. *Curr Pharmacol Rep* 5(4): 255–80.
Al-Yahya, A.A., A.A. Al-Majed, A.M. Gado, M.H. Daba, O.A. Al-Shabanah, and A.R. Abd-Allah. 2009. *Acacia Senegal* gum exudate offers protection against cyclophosphamide-induced urinary bladder cytotoxicity. *Oxid Med Cell Longev* 2(4): 207–13.
Alajmi, M.F., P. Alam, S.I. Alqasoumi, et al. 2017. Comparative anticancer and antimicrobial activity of aerial parts of *Acacia salicina*, *Acacia laeta*, *Acacia hamulosa* and *Acacia tortilis* grown in Saudi Arabia. *Saudi Pharm J* 25(8): 1248–52.
Anon. 2005. Final report of the safety assessment of *Acacia catechu* gum, *Acacia concinna* fruit extract, *Acacia dealbata* leaf extract, *Acacia dealbata* leaf wax, *Acacia decurrens* extract, *Acacia farnesiana* extract, *Acacia farnesiana* flower wax, *Acacia farnesiana* gum, *Acacia senegal* extract, *Acacia senegal* gum, and *Acacia senegal* gum extract. *Int J Toxicol* 24(Suppl 3): 75–118.
Aruna, C. 2014. Preliminary phytochemical screening of *Acacia caesia* (L.) Willd. *IJPBS* 5(2): B484–B490.
Auwal, M.S., S. Saka, I.A. Mairiga, K.A. Sanda, A. Shuaibu, and A. Ibrahim. 2014. Preliminary phytochemical and elemental analysis of aqueous and fractionated pod extracts of *Acacia nilotica* (Thorn mimosa). *Vet Res Forum* 5(2): 95–100.
Aydar, E., D. Stratton, S.P. Fraser, M.B. Djamgoz, and C. Palmer. 2016. Sigma-1 receptors modulate neonatal $Na_v1.5$ ion channels in breast cancer cell lines. *Eur Biophys J* 45(7): 671–83.
Bachran, C., S. Bachran, M. Sutherland, D. Bachran, and H. Fuchs. 2008. Saponins in tumor therapy. *Mini Rev Med Chem* 8(6): 575–84.
Baig, S., M.H. Lucky, A. Qamar, et al. 2012. Human papilloma virus and oral lesions in gutka eating subjects in Karachi. *J Coll Physicians Surg Pak* 22(3): 135–8.
Barapatre, A., A.S. Meena, S. Mekala, A. Das, and H. Jha. 2016. *In vitro* evaluation of antioxidant and cytotoxic activities of lignin fractions extracted from *Acacia nilotica*. *Int J Biol Macromol* 86: 443–53.
Buchanan, M.S., A.R. Carroll, D. Pass, and R.J. Quinn. 2007. NMR spectral assignments of a new chlorotryptamine alkaloid and its analogues from *Acacia confusa*. *Magn Reson Chem* 45(4): 359–61.
Burnett, B.P., Q. Jia, Y. Zhao, and R.M. Levy. 2007. A medicinal extract of *Scutellaria baicalensis* and *Acacia catechu* acts as a dual inhibitor of cyclooxygenase and 5-lipoxygenase to reduce inflammation. *J Med Food* 10(3): 442–51.
Chatti, I.B., I. Limem, J. Boubaker, et al. 2009. Phytochemical, antibacterial, antiproliferative, and antioxidant potentials and DNA damage-protecting activity of *Acacia salicina* extracts. *J Med Food* 12(3): 675–83.
Chiaino, E., M. Micucci, M. Durante, et al. 2020. Apoptotic-induced effects of *Acacia catechu* Willd. extract in human colon cancer cells. *Int J Mol Sci* 21(6): 2102.
Chundakkattumalayil, H.C., S. Kumar, R. Narayanan, and K.T. Raghavan. 2019. Role of *L. plantarum* KX519413 as probiotic and acacia gum as prebiotic in gastrointestinal tract strengthening. *Microorganisms* 7(12): 659.
Collina, S., E. Bignardi, M. Rui, et al. 2017. Are sigma modulators an effective opportunity for cancer treatment? A patent overview (1996–2016). *Expert Opin Ther Pat* 27(5): 565–78.
Croce, C.M. 2001. How can we prevent cancer? *Proc Natl Acad Sci U S A* 98(20): 10986–8.
Diab, K.A., S.K. Guru, S. Bhushan, and A.K. Saxena. 2015. *In vitro* anticancer activities of *Anogeissus latifolia*, *Terminalia bellerica*, *Acacia catechu* and *Moringa oleifera* Indian plants. *Asian Pac J Cancer Prev* 16(15): 6423–8.
De Nicola, E., M. Gallo, M. Iaccarino, et al. 2004. Hormetic versus toxic effects of vegetable tannin in a multitest study. *Arch Environ Contam Toxicol* 46(3): 336–44.
Deiab, S., E. Mazzio, S. Messeha, N. Mack, and K.F. Soliman. 2013. High-throughput screening to identify plant derived human LDH-A inhibitors. *European J Med Plants* 3(4): 603–15.
Dylan, B. 1989. The disease of conceit. In *Oh Mercy*. New York, NY: Columbia Records.
Fotso, W.G., J. Na-Iya, T.A. Mbaveng, et al. 2019. Polyacanthoside A, a new oleanane-type triterpenoid saponin with cytotoxic effects from the leaves of *Acacia polyacantha* (Fabaceae). *Nat Prod Res* 33(24): 3521–6.
Gaikwad, A., A. Poblenz, V. Haridas, C. Zhang, M. Duvic, and J. Gutterman. 2005. Triterpenoid electrophiles (avicins) suppress heat shock protein-70 and X-linked inhibitor of apoptosis proteins in malignant cells by activation of ubiquitin machinery: Implications for proapoptotic activity. *Clin Cancer Res* 11(5): 1953–62.

Gamble, J.S. 1902. *A Manual of Indian Timbers: An Account of the Growth, Distribution, and Uses of the Trees and Shrubs of India and Ceylon with Descriptions of Their Wood-Structure*. London: S. Low, Marston & Co.

Gedara, S.R., and A.A. Galala. 2014. New cytotoxic spirostane saponin and biflavonoid glycoside from the leaves of *Acacia saligna* (Labill.) H.L. Wendl. *Nat Prod Res* 28(5): 324–9.

Georgiadis, M.O., O. Karoutzou, A.S. Foscolos, and I. Papanastasiou. 2017. Sigma receptor (σR) ligands with antiproliferative and anticancer activity. *Molecules* 22(9): 1408.

Ghate, N.B., B. Hazra, R. Sarkar, and N. Mandal. 2014. Heartwood extract of *Acacia catechu* induces apoptosis in human breast carcinoma by altering bax/bcl-2 ratio. *Pharmacogn Mag* 10(37): 27–33.

Gomes, M.M., J.B. Coimbra, R.O. Clara, et al. 2014. A biosynthesis of N,N-dimethyltryptamine (DMT) in a melanoma cell line and its metabolization by peroxidases. *Biochem Pharmacol* 88: 393–401.

Grace, M.H., G.R. Wilson, F.E. Kandil, E. Dimitriadis, and R.M. Coates. 2009. Characteristic flavonoids from *Acacia burkittii* and *A. acuminata* heartwoods and their differential cytotoxicity to normal and leukemia cells. *Nat Prod Commun* 4(1): 69–76.

Gulati, V., P. Gulati, I.H. Harding, and E.A. Palombo. 2015. Exploring the anti-diabetic potential of Australian Aboriginal and Indian Ayurvedic plant extracts using cell-based assays. *BMC Complement Altern Med* 15: 8.

Gutterman, J.U., H.T. Lai, P. Yang, V. Haridas, A. Gaikwad, and S. Marcus. 2005. Effects of the tumor inhibitory triterpenoid avicin G on cell integrity, cytokinesis, and protein ubiquitination in fission yeast. *Proc Natl Acad Sci U S A* 102(36): 12771–6.

Hanausek, M., P. Ganesh, Z. Walaszek, C.J. Arntzen, T.J. Slaga, and J.U. Gutterman. 2001. Avicins, a family of triterpenoid saponins from *Acacia victoriae* (Bentham), suppress H-ras mutations and aneuploidy in a murine skin carcinogenesis model. *Proc Natl Acad Sci U S A* 98(20): 11551–6.

Haridas, V., C.J. Arntzen, and J.U. Gutterman. 2001a. Avicins, a family of triterpenoid saponins from *Acacia victoriae* (Bentham), inhibit activation of nuclear factor-kappaB by inhibiting both its nuclear localization and ability to bind DNA. *Proc Natl Acad Sci U S A* 98(20): 11557–62.

Haridas, V., M. Hanausek, G. Nishimura, et al. 2004. Triterpenoid electrophiles (avicins) activate the innate stress response by redox regulation of a gene battery. *J Clin Invest* 113(1): 65–73.

Haridas, V., M. Higuchi, G.S. Jayatilake, et al. 2001b. Avicins: Triterpenoid saponins from *Acacia victoriae* (Bentham) induce apoptosis by mitochondrial perturbation. *Proc Natl Acad Sci U S A* 98(10): 5821–6.

Haridas, V., X. Li, T. Mizumachi, et al. 2007. Avicins, a novel plant-derived metabolite lowers energy metabolism in tumor cells by targeting the outer mitochondrial membrane. *Mitochondrion* 7(3): 234–40.

Haridas, V., G. Nishimura, Z.X. Xu, et al. 2009. Avicin D: A protein reactive plant isoprenoid dephosphorylates Stat 3 by regulating both kinase and phosphatase activities. *PLoS One* 4(5): e5578.

Haridas, V., Z.X. Xu, D. Kitchen, A. Jiang, P. Michels, and J.U. Gutterman. 2011. The anticancer plant triterpenoid, avicin D, regulates glucocorticoid receptor signaling: Implications for cellular metabolism. *PLoS One* 6(11): e28037.

Hemachandran, H., A. Anantharaman, R.R. Priya, G.P. Doss, and R. Siva. 2016. Interaction of catechu dye with DNA: Spectroscopic and in silico approach. *Nucleos Nucleot Nucl* 35(4): 195–210.

Huang, W., H. Niu, X. Xue, J. Li, and C. Li. 2010. Robinetinidol-(4beta-->8)-epigallocatechin 3-O-gallate, a galloyl dimer prorobinetinidin from *Acacia mearnsii* De Wild, effectively protects human neuroblastoma SH-SY5Y cells against acrolein-induced oxidative damage. *J Alzheimers Dis* 21(2): 493–506.

Huang, Y.S., H.L. Lu, L.J. Zhang, and Z. Wu. 2014. Sigma-2 receptor ligands and their perspectives in cancer diagnosis and therapy. *Med Res Rev* 34(3): 532–66.

Jæger, D., C.P. Ndi, C. Crocoll, et al. 2017. Isolation and structural characterization of echinocystic acid triterpenoid saponins from the Australian medicinal and food plant *Acacia ligulata*. *J Nat Prod* 80(10): 2692–8.

Jæger, D., B.S. Simpson, C.P. Ndi, et al. 2018. Biological activity and LC-MS/MS profiling of extracts from the Australian medicinal plant *Acacia ligulata* (Fabaceae). *Nat Prod Res* 32(5): 576–81.

Javed, F., M. Chotai, A. Mehmood, and K. Almas. 2010. Oral mucosal disorders associated with habitual gutka usage: A review. *Oral Surg Oral Med Oral Pathol Oral Radiol Endod* 109(6): 857–64.

Jelassi, A., A. Zardi-Bergaoui, A. Ben Nejma, M. Belaiba, J. Bouajila, and H. Ben Jannet. 2014. Two new unusual monoterpene acid glycosides from *Acacia cyclops* with potential cytotoxic activity. *Bioorg Med Chem Lett* 24(16): 3777–81.

Joshi, L., J.M. Van Eck, K. Mayo, et al. 2002. Metabolomics of plant saponins: Bioprospecting triterpene glycoside diversity with respect to mammalian cell targets. *OMICS* 6(3): 235–46.

Kalaivani, T., C. Rajasekaran, K. Suthindhiran, and L. Mathew. 2011. Free radical scavenging, cytotoxic and hemolytic activities from leaves of *Acacia nilotica* (L.) Wild. ex. Delile subsp. *indica* (Benth.) Brenan. *Evid Based Complement Alternat Med* 2011: 274741.

Kar, I., and R. Chattopadhyaya. 2017. Effect of seven Indian plant extracts on Fenton reaction-mediated damage to DNA constituents. *J Biomol Struct Dyn* 35(14): 2997–3011.

Kaur, K., S. Arora, M.E. Hawthorne, S. Kaur, S. Kumar, and R.G. Mehta. 2002. A correlative study on antimutagenic and chemopreventive activity of *Acacia auriculiformis* A. Cunn. and *Acacia nilotica* (L.) Willd. ex Del. *Drug Chem Toxicol* 25(1): 39–64.

Lakshmi, T., D. Ezhilarasan, U. Nagaich, and R. Vijayaragavan. 2017a. *Acacia catechu* ethanolic seed extract triggers apoptosis of SCC-25 cells. *Pharmacogn Mag* 13(Suppl 3): S405–S411.

Lakshmi, T., D. Ezhilarasan, R. Vijayaragavan, S.K. Bhullar, and R. Rajendran. 2017b. *Acacia catechu* ethanolic bark extract induces apoptosis in human oral squamous carcinoma cells. *J Adv Pharm Technol Res* 8(4): 143–9.

Lam, S.K., and T.B. Ng. 2010a. A dimeric high-molecular-weight chymotrypsin inhibitor with antitumor and HIV-1 reverse transcriptase inhibitory activities from seeds of *Acacia confusa*. *Phytomedicine* 17(8–9): 621–5.

Lam, S.K., and T.B. Ng. 2010b. Acaconin, a chitinase-like antifungal protein with cytotoxic and anti-HIV-1 reverse transcriptase activities from *Acacia confusa* seeds. *Acta Biochim Pol* 57(3): 299–304.

Lansky, E.P., G. Harrison, P. Froom, and W.G. Jiang. 2005. Pomegranate (*Punica granatum*) pure chemicals show possible synergistic inhibition of human PC-3 prostate cancer cell invasion across Matrigel [correction published in 2005 in *Invest New Drugs* 23(4): 379]. *Invest New Drugs* 23(2): 121–2.

Lemeshko, V.V., V. Haridas, J.C. Quijano Pérez, and J.U. Gutterman. 2006. Avicins, natural anticancer saponins, permeabilize mitochondrial membranes. *Arch Biochem Biophys* 454(2): 114–22.

Lin, A.S., C.R. Lin, Y.C. Du, et al. 2009. Acasiane A and B and farnesirane A and B, diterpene derivatives from the roots of *Acacia farnesiana*. *Planta Med* 75(3): 256–61.

Lin, J.Y., Y.S. Hsieh, and S.C. Chu. 1989. Chimeric protein: Abrin B chain-trypsin inhibitor conjugate as a new antitumor agent. *Biochem Int* 19(2): 313–23.

Lin, J.Y., and L.L. Lin. 1985. Antitumor lectin-trypsin inhibitor conjugate. *J Natl Cancer Inst* 74(5): 1031–6.

Liu, C.H., W.L. Chu, S.C. Liao, C.C. Yang, and C.C. Lin. 2019. Syrian rue seeds interacted with acacia tree bark in an herbal stew resulted in N,N-dimethyltryptamine poisoning. *Clin Toxicol (Phila)* 57(10): 867–9.

Man, S., W. Gao, Y. Zhang, L. Huang, and C. Liu. 2010. Chemical study and medical application of saponins as anti-cancer agents. *Fitoterapia* 81(7): 703–14.

Mazzio, E.A., and K.F. Soliman. 2017. HTP nutraceutical screening for histone deacetylase inhibitors and effects of HDACis on tumor-suppressing miRNAs by Trichostatin A and grapeseed (*Vitis vinifera*) in HeLa cells. *Cancer Genomics Proteomics* 14(1): 17–33.

Meena, P.D., P. Kaushik, S. Shukla, A.K. Soni, M. Kumar, and A. Kumar. 2006. Anticancer and antimutagenic properties of *Acacia nilotica* (Linn.) on 7,12-dimethylbenz(a)anthracene-induced skin papillomagenesis in Swiss albino mice. *Asian Pac J Cancer Prev* 7(4): 627–32.

Mishra, G.A., S.S. Gunjal, S.A. Pimple, P.V. Majmudar, S.D. Gupta, and S.S. Shastri. 2014. Impact of 'gutkha and pan masala ban' in the state of Maharashtra on users and vendors. *Indian J Cancer* 51(2): 129–32.

Mohammed, A.E., A. Al-Qahtani, A. Al-Mutairi, B. Al-Shamri, and K.F. Aabed. 2018. Antibacterial and cytotoxic potential of biosynthesized silver nanoparticles by some plant extracts. *Nanomaterials (Basel)* 8(6): 382.

Mohan, S., K. Thiagarajan, and R. Chandrasekaran. 2017. Evaluation of ethyl gallate for its antioxidant and anticancer properties against chemical-induced tongue carcinogenesis in mice. *Biochem J* 474(17): 3011–25.

Mohan, S., K. Thiagarajan, R. Chandrasekaran, and J. Arul. 2014. *In vitro* protection of biological macromolecules against oxidative stress and *in vivo* toxicity evaluation of *Acacia nilotica* (L.) and ethyl gallate in rats. *BMC Complement Altern Med* 14: 257.

Mollica, A., M. Locatelli, A. Stefanucci, and F. Pinnen. 2012. Synthesis and bioactivity of secondary metabolites from marine sponges containing dibrominated indolic systems. *Molecules* 17(5): 6083–99.

Monga, J., C.S. Chauhan, and M. Sharma. 2013. Human breast adenocarcinoma cytotoxicity and modulation of 7,12-dimethylbenz[a]anthracene-induced mammary carcinoma in Balb/c mice by *Acacia catechu* (L.f.) Wild heartwood. *Integr Cancer Ther* 12(4): 347–62.

Morton, J.F. 1992. Widespread tannin intake via stimulants and masticatories, especially guarana, kola nut, betel vine, and accessories. *Basic Life Sci* 59: 739–65.

Mothana, R.A., U. Lindequist, R. Gruenert, and P.J. Bednarski. 2009. Studies of the in vitro anticancer, antimicrobial and antioxidant potentials of selected Yemeni medicinal plants from the island Soqotra. *BMC Complement Altern Med* 9: 7.

Mujoo, K., V. Haridas, J.J. Hoffmann, et al. 2001. Triterpenoid saponins from *Acacia victoriae* (Bentham) decrease tumor cell proliferation and induce apoptosis. *Cancer Res* 61(14): 5486–90.
Mutai, C., D. Abatis, C. Vagias, D. Moreau, C. Roussakis, and V. Roussis. 2007. Lupane triterpenoids from *Acacia mellifera* with cytotoxic activity. *Molecules* 12(5): 1035–44.
Nasir, O. 2013. Renal and extrarenal effects of gum arabic (*Acacia senegal*)—what can be learned from animal experiments? *Kidney Blood Press Res* 37(4–5): 269–79.
Nasir, O., K. Wang, M. Föller, et al. 2010. Downregulation of angiogenin transcript levels and inhibition of colonic carcinoma by gum arabic (*Acacia senegal*). *Nutr Cancer* 62(6): 802–10.
Pagano, G., G. Castello, M. Gallo, I. Borriello, and M. Guida. 2008. Complex mixture-associated hormesis and toxicity: The case of leather tanning industry. *Dose Response* 6(4): 383–96.
Pal, K., S. Roy, P.K. Parida, et al. 2019. Folic acid conjugated curcumin loaded biopolymeric gum acacia microsphere for triple negative breast cancer therapy in in vitro and in vivo model. *Mater Sci Eng C Mater Biol Appl* 95: 204–16.
Pal, K., S. Roy, P.K. Parida, et al. 2020. Corrigendum to "Folic acid conjugated curcumin loaded biopolymeric gum acacia microsphere for triple negative breast cancer therapy in in vitro and in vivo model." *Mater Sci Eng C Mater Biol Appl* 111: 110866.
Popoca, J., A. Aguilar, D. Alonso, and M.L. Villarreal. 1998. Cytotoxic activity of selected plants used as antitumorals in Mexican traditional medicine. *J Ethnopharmacol* 59(3): 173–7.
Qiu, M., J. Liu, Y. Su, R. Guo, B. Zhao, and J. Liu. 2020. Diosmetin induces apoptosis by downregulating AKT phosphorylation via P53 activation in human renal carcinoma ACHN cells. *Protein Pept Lett* 27(10): 1022–8.
Rao, T.N., Riyazuddin, P. Babji, et al. 2019. Green synthesis and structural classification of *Acacia nilotica* mediated-silver doped titanium oxide (Ag/TiO_2) spherical nanoparticles: Assessment of its antimicrobial and anticancer activity. *Saudi J Biol Sci* 26(7): 1385–91.
Revathi, S., F.L. Hakkim, N.R. Kumar, et al. 2018. Induction of HT-29 colon cancer cells apoptosis by pyrogallol with growth inhibiting efficacy against drug-resistant *Helicobacter pylori*. *Anticancer Agents Med Chem* 18(13): 1875–84.
Rifai, Y., M.A. Arai, T. Koyano, T. Kowithayakorn, and M. Ishibashi. 2010. Terpenoids and a flavonoid glycoside from *Acacia pennata* leaves as Hedgehog/GLI-mediated transcriptional inhibitors. *J Nat Prod* 73(5): 995–7.
Sakthivel, K.M., and C. Guruvayoorappan. 2013. *Acacia ferruginea* inhibits tumor progression by regulating inflammatory mediators—(TNF-α, iNOS, COX-2, IL-1β, IL-6, IFN-γ, IL-2, GM-CSF) and pro-angiogenic growth factor—VEGF. *Asian Pac J Cancer Prev* 14(6): 3909–19.
Sakthivel, K.M., and C. Guruvayoorappan. 2018. Targeted inhibition of tumor survival, metastasis and angiogenesis by *Acacia ferruginea* mediated regulation of VEGF, inflammatory mediators, cytokine profile and inhibition of transcription factor activation. *Regul Toxicol Pharmacol* 95: 400–11.
Sakthivel, K.M., N. Kannan, A. Angeline, and C. Guruvayoorappan. 2012. Anticancer activity of *Acacia nilotica* (L.) Wild. Ex. Delile subsp. *indica* against Dalton's ascitic lymphoma induced solid and ascitic tumor model. *Asian Pac J Cancer Prev* 13(8): 3989–95.
Sawadogo, W.R., A. Maciuk, J.T. Banzouzi, et al. 2012. Mutagenic effect, antioxidant and anticancer activities of six medicinal plants from Burkina Faso. *Nat Prod Res* 26(6): 575–9.
Seo, Y., J. Hoch, M. Abdel-Kader, et al. 2002. Bioactive saponins from *Acacia tenuifolia* from the Suriname Rainforest. *J Nat Prod* 65(2): 170–4.
Sharma, V., A. Nandan, A.K. Shukla, et al. 2018. *Dohra*—A mixture of potent carcinogens. *Indian J Med Res* 148(1): 116–19.
Shen, X., Y. Wang, and F. Wang. 2010. Characterisation and biological activities of proanthocyanidins from the barks of *Pinus massonian* and *Acacia mearnsii*. *Nat Prod Res* 24(6): 590–8.
Singh, B.N., B.R. Singh, B.K. Sarma, and H.B. Singh. 2009. Potential chemoprevention of N-nitrosodiethylamine-induced hepatocarcinogenesis by polyphenolics from *Acacia nilotica* bark. *Chem Biol Interact* 181(1): 20–28.
Soriani, O., and R. Rapetti-Mauss. 2017. Sigma 1 receptor and ion channel dynamics in cancer. *Adv Exp Med Biol* 964: 63–77.
Sundarraj, S., R. Thangam, V. Sreevani, et al. 2012. γ-Sitosterol from *Acacia nilotica* L. induces G2/M cell cycle arrest and apoptosis through c-Myc suppression in MCF-7 and A549 cells. *J Ethnopharmacol* 141(3): 803–9.
Szabo, A. 2015. Psychedelics and immunomodulation: Novel approaches and therapeutic opportunities. *Front Immunol* 6: 358.

Torres-Valencia, J.M., V. Motilva, J.J. Manríquez-Torres, et al. 2015. Antiproliferative activity of seco-oxacassanes from *Acacia schaffneri*. *Nat Prod Commun* 10(6): 853–6.

Tourino, M.C., E.M. de Oliveira, L.P. Bellé, et al. 2013. Tryptamine and dimethyltryptamine inhibit indoleamine 2,3 dioxygenase and increase the tumor-reactive effect of peripheral blood mononuclear cells. *Cell Biochem Funct* 31(5): 361–4.

Tsammalex: A lexical database on plants and animals. 2020. Ed. by Ch. Naumann, T. Güldemann, S. Moran, G. Segerer, A.-M. Fehn, and R. Forkel. Leipzig: Max Planck Institute for Evolutionary Anthropology. http://tsammalex.clld.org [accessed on 23/8/2022].

Vadivel, V., and H.K. Biesalski. 2012. Effect of certain indigenous processing methods on the bioactive compounds of ten different wild type legume grains. *J Food Sci Technol* 49(6): 673–84.

van Elswijk, D.A., U.P. Schobel, E.P. Lansky, H. Irth, and J. van der Greef. 2004. Rapid dereplication of estrogenic compounds in pomegranate (*Punica granatum*) using on-line biochemical detection coupled to mass spectrometry. *Phytochemistry* 65(2): 233–41.

van Waarde, A., A.A. Rybczynska, N. Ramakrishnan, K. Ishiwata, P.H. Elsinga, and R.A. Dierckx. 2010. Sigma receptors in oncology: Therapeutic and diagnostic applications of sigma ligands. *Curr Pharm Des* 16(31): 3519–37.

van Waarde, A., A.A. Rybczynska, N.K. Ramakrishnan, K. Ishiwata, P.H. Elsinga, and R.A. Dierckx. 2015. Potential applications for sigma receptor ligands in cancer diagnosis and therapy. *Biochim Biophys Acta* 1848(10 Pt B): 2703–14.

White, E.P. 1957. Alkaloids of the Leguminosae XXVI: Examination of further legumes, mainly *Lupinus* and *Acacia* species for alkaloids. *N Z J Sci Technol Sect B* 38: 718–25.

Xu, Z.X., T. Ding, V. Haridas, F. Connolly, and J.U. Gutterman. 2009. Avicin D, a plant triterpenoid, induces cell apoptosis by recruitment of Fas and downstream signaling molecules into lipid rafts. *PLoS One* 4(12): e8532.

Xu, Z.X., J. Liang, V. Haridas, et al. 2007. A plant triterpenoid, avicin D, induces autophagy by activation of AMP-activated protein kinase. *Cell Death Differ* 14(11): 1948–57.

Yadav, V.R., S. Prasad, B. Sung, R. Kannappan, and B.B. Aggarwal. 2010. Targeting inflammatory pathways by triterpenoids for prevention and treatment of cancer. *Toxins (Basel)* 2(10): 2428–66.

Yan, J.J., R.W. Sun, P. Wu, M.C. Lin, A.S. Chan, and C.M. Che. 2010. Encapsulation of dual cytotoxic and anti-angiogenic gold(III) complexes by gelatin-acacia microcapsules: *In vitro* and *in vivo* studies. *Dalton Trans* 39(33): 7700–5.

Zhang, C., B. Li, A.S. Gaikwad, et al. 2008. Avicin D selectively induces apoptosis and downregulates p-STAT-3, bcl-2, and survivin in cutaneous T-cell lymphoma cells. *J Invest Dermatol* 128(11): 2728–35.

Zingue, S., T. Michel, J. Cisilotto, et al. 2018a. The hydro-ethanolic extract of *Acacia seyal* (Mimosaceae) stem barks induced death in an ER-negative breast cancer cell line by the intrinsic pathway of apoptosis and inhibited cell migration. *J Ethnopharmacol* 223: 41–50.

Zingue, S., A.N. Njuh, A.B. Tueche, et al. 2018b. *In vitro* cytotoxicity and *in vivo* antimammary tumor effects of the hydroethanolic extract of *Acacia seyal* (Mimosaceae) stem bark. *Biomed Res Int* 2018: 2024602.

6 Psychotropic Acacia

It is hard to say exactly what percentage of the world drug market is given to psychoactive substances, but one can be sure it is very high and constantly growing, considering the leakage of such drugs against anxiety and depression from the hands of psychiatrists to those of general practitioners and family medicine specialists. Taking into account drinks with caffeine, like coffee, it seems as if nearly everyone uses psychoactive plant products.

The word *psychotropic* means literally a turning toward the mind or an affinity for it, in short, drugs that affect the mind. These include medicines for the mood, such as tranquilizers, stimulants, antidepressants, and anxiolytics, which may also be applied for learning disorders, physical pain, epilepsy, and many other related indications. All of these drugs including coffee (*Coffea* sp.), tea (*Thea* sp.), or chocolate (*Theobroma cacao*), are psychotropic.

Psychotropic research on plant parts from *Acacia s.l.* is surprisingly scanty, although the implications of some of it are staggering. Responding to the use of plant parts of *Acacia s.l.* in Zimbabwe for treatment of convulsions, a Saudi team took leaves of *Vachellia tortilis* and blended them with distilled water, filtering the product to obtain the extract. The extract was then administered to mice to evaluate potential anti-seizure activity. The *V. tortilis* extract at 400 or 800 mg/kg prolonged seizure latency (time to onset) in mice, with the change in latency in time being superior to that produced by the classic fast-acting anticonvulsant, the benzodiazepine diazepam (Alharbi and Azmat 2015). In a second study with the same extract at the same dosages, the authors showed anxiolytic effects of the extract in mice at low dosages and sedative effects at high dosages, and they suggest that the effect was attributable to dimethyltryptamine (DMT) contained within the aqueous extract (Alharbi and Azmat 2016).

The esteemed Professor Afsar and her team showed their methanolic and ethyl acetate extracts of the aerial parts (bark, twigs, and leaves) of *Acacia eburnea* (*A. hydaspica*) to have an improving effect in mice *in vivo* for climbing behavior, an index of mood, comparable to that achieved with the antidepressant fluoxetine, and an anxiolytic effect in the elevated plus-maze test (EPM) comparable to that of diazepam. Compounds identified by GC-MS in the extracts included *1,2-benzenedicarboxylic acid mono (2-ethylhexyl) ester* (70.65%), *α-amyrin* (5.03%), *vitamin E* (4.56%), *2,6-dimethyl-N-(2-methyl-α-phenylbenzyl) aniline* (2.51%) and *squalene* (4%) (Afsar et al. 2017).

As for potentially psychotropic *Acacia s.l.* phytochemicals, an amphetamine isomer, *N-methyl beta-phenylethylamine* (NMBP), was isolated from *Acacia berlandieri* (Camp and Lyman 1956), along with *tyramine* and *N-methyl tyramine* (Camp et al. 1964). Later, NMBP was also isolated from *Acacia rigidula* (Pawar et al. 2014). The pure NMBP has been the focus of hullabaloo regarding its presence in certain dietary supplements that boast "boosting" properties (Rickli et al. 2019; Yun et al. 2017).

Using reverse-phase high-pressure liquid chromatography, researchers studied an extract of the leaves of *Acacia podalyriifolia*, succeeding to isolate *tryptamine* from the extract and *N,N-dimethyltryptamine* from urine after consumption (Balandrin et al. 1978). A methanolic extract of *Acacia confusa* was used to find *N,N-dimethyltryptamine*, *N-methyltryptamine*, *N,N-dimethyltryptamine-N-oxide*, and *N-chloromethyl-N,N-dimethyltryptamine* (Buchanan et al. 2007).

Dimethyltryptamine (DMT) in particular and substituted tryptamines in general are known for producing fabulous hallucinogenic, psychedelic, and entheogenic effects in receptive individuals, so long as certain conditions are met. One major part of the so-called blood-brain barrier is the enzyme known as monoamine oxidase (MAO), which is on constant patrol against compounds like DMT and, if allowed, will oxidize those substituted tryptamines into so many powerless metabolites. One trick is to put a heavy phosphate group on the molecule, as *psilocin*

DOI: 10.1201/9780429440946-6

FIGURE 6.1 The beautiful "pearl acacia" or "Queensland silver wattle," *Acacia podalyriifolia*, is a fast-growing tree, indigenous to Australia, transplanted to India, Malaysia, Africa, and South America. It is a perennial, growing to 5 m in height, and appreciated mainly as an ornamental. (a) *A. podalyriifolia* buds, flowers, and leaves. (12/11/2020, San Diego Botanic Garden, Hamilton Children's Garden, Encinitas, CA: by K M, CC BY 2.0, via Flickr, www.flickr.com/photos/131880272@N06/50784503387.) (b) *A. podalyriifolia* branches with seed pods. (4/15/2005, Kula, Maui, Hawaii: by Forest & Kim Starr, CC BY 4.0, via Starr Environmental, www.starrenvironmental.com/images/image/?q=24719081856.)

did to become *psilocybin*, which seems to obstruct the oxidative enzymes, at least for a while. Another hack is to mix the DMT-containing biomass with some other natural material that contains inhibitors of MAO, such as the harmala alkaloids *harmine*, *harmaline*, and *tetrahydroharmine* found in the seeds of *Peganum harmala* (Lansky et al. 2017). This was the strategy apparently adopted by the college roommates in Professor Liu's case report, in which it was claimed that the violent behavioral reaction was a result of "dimethyltryptamine poisoning" (Liu et al. 2019).

Regardless of the technicalities of what enabled the "poisoning" to occur, a larger question still looms: what led these young men, or any other seemingly sane, thinking person, to intentionally consume an herbal concoction that is likely to induce, at least temporarily, a major change, or shift, in consciousness? Why did they do it? Especially when, as in the case of the South American brew, *ayahuasca*, which possesses a similar chemistry to *A. confusa* plus *P. harmala*, drinking is often accompanied by intense and sometimes violent vomiting? For what purpose would a sane person do such a thing? For the "fun" of hallucinating?

In a most amazing treatise, P.D. Newman, a 32nd-degree Master Mason, concludes that the ancients were not only aware of the psychoactive properties of acacias through the "philosopher's stone," *i.e.*, DMT, which they contained (exemplified by a sprig of acacia as a centuries-old symbol adorning Masonic documents and paraphernalia) but intentionally cultivated acacia use to an elite whose charge it was to renew and maintain the mystical tradition that its entheogenic employment had originally kindled (Newman 2017). Used by itself and injected intravenously to bypass the MAO in the digestive tract, DMT is among the most potent hallucinogens known, facilitating not only the most spectacular of visions but frank phenomenological transport to parallel dimensions for consort with the non-human beings that seem to inhabit such realms (Strassman 2001).

Interdimensional beings and masonic initiations are well and good, but how is all this relevant to the present discussion among moderns with a scientific, even medically centered, outlook? Plenty! Dimethyltryptamine as the primary vision-producing agent in the South American entheogenic brew ayahuasca or in many "analogues" employing other plants for comparable pharmacological effects (Ott 1994) has, it turns out, startling ramifications not only for psychiatry, for example in the treatment of addictions (Hamill et al. 2019), depression (Cameron et al. 2018, 2019), and anxiety (Davis et al. 2019), but for somatic medicine as well through the agonism of the sigma-1 receptor by DMT (it is a perfect ligand on structural grounds). From this agonism originates DMT's potential utility for ameliorating visual impairment secondary to retinal deterioration (along with companion plants with temporary MAO-inhibiting properties, such as species from the genera *Banisteriopsis* or *Peganum* (Lansky 2020b), for the treatment of cancer (Schenberg 2013; van Waarde et al. 2015), and for benefiting other treatment-resistant general medical or "psychosomatic" maladies (Maurice and Su 2009; Szabo 2015). Similarly, DMT may modulate damage from stroke (Nardai et al. 2020). On a psychic level, DMT, particularly when used in the context of an Ayahuasca "ceremony," has the potential to enable a person to heal from addictions by reconnecting dismembered experiences to consciousness in the process of "remembering one's self" (Inserra 2018; Lansky 2020a; Mate 2017; Tafur 2017).

Last, one final curiosity concerning DMT and other substituted tryptamine psychedelics is their capacity to initiate clinical changes that can persist for weeks, months, or even years after the therapeutic dose (de Almeida et al. 2019; Kuypers 2019; Sanches et al. 2016). Why this works is not entirely clear, though possibly it could relate to a kind of re-tuning of serotonergic receptors, modulating their individual receptivities to particular ligands under particular conditions (Lansky et al. 2017). If this is correct, it would imply a completely different therapeutic approach, one based on infrequent doses with long-term consequences rather than repeated daily or more frequent dosing with diminishing effectiveness and accumulating toxicity. The new style would be environmental- and patient-friendly as well as economical.

REFERENCES

Afsar, T., S. Razak, M.R. Khan, and A. Almajwal. 2017. Anti-depressant and anxiolytic potential of *Acacia hydaspica* R. Parker aerial parts extract: Modulation of brain antioxidant enzyme status. *BMC Complement Altern Med* 17(1): 228.

Alharbi, W.D.M., and A. Azmat. 2015. Anticonvulsant and neuroprotective effects of the *Acacia tortilis* growing in KSA. *Pak J Pharm Sci* 28(2): 531–4.

Alharbi, W.D.M., and A. Azmat. 2016. Pharmacological evidence of neuro-pharmacological activity of *Acacia tortilis* leaves in mice. *Metab Brain Dis* 31(4): 881–5.

Balandrin, M.F., A.D. Kinghorn, S.J. Smolenski, and R.H. Dobberstein. 1978. Reversed-phase high-pressure liquid chromatography of some tryptamine derivatives. *J Chromatogr* 157: 365–70.

Buchanan, M.S., A.R. Carroll, D. Pass, and R.J. Quinn. 2007. NMR spectral assignments of a new chlorotryptamine alkaloid and its analogues from *Acacia confusa*. *Magn Reson Chem* 45(4): 359–61.

Cameron, L.P., C.J. Benson, B.C. DeFelice, O. Fiehn, and D.E. Olson. 2019. Chronic, intermittent microdoses of the psychedelic *N,N*-dimethyltryptamine (DMT) produce positive effects on mood and anxiety in rodents. *ACS Chem Neurosci* 10(7): 3261–70.

Cameron, L.P., C.J. Benson, L.E. Dunlap, and D.E. Olson. 2018. Effects of N, N-dimethyltryptamine on rat behaviors relevant to anxiety and depression. *ACS Chem Neurosci* 9(7): 1582–90.

Camp, B.J., R. Adams, and J.W. Dollahite. 1964. The chemistry of the toxic constituents of *Acacia berlandieri*. *Ann N Y Acad Sci* 111: 744–50.

Camp, B.J., and C.M. Lyman. 1956. The isolation of N-methyl beta-phenylethylamine from *Acacia berlandieri*. *J Am Pharm Assoc Am Pharm Assoc* 45(11): 719–21.

Davis, A.K., S. So, R. Lancelotta, J.P. Barsuglia, and R.R. Griffiths. 2019. 5-methoxy-N,N-dimethyltryptamine (5-MeO-DMT) used in a naturalistic group setting is associated with unintended improvements in depression and anxiety. *Am J Drug Alcohol Abuse* 45(2): 161–9.

de Almeida, R.N., A.C.M. Galvão, F.S. da Silva, et al. 2019. Modulation of serum brain-derived neurotrophic factor by a single dose of ayahuasca: Observation from a randomized controlled trial. *Front Psychol* 10: 1234.

Hamill, J., J. Hallak, S.M. Dursun, and G. Baker. 2019. Ayahuasca: Psychological and physiologic effects, pharmacology and potential uses in addiction and mental illness. *Curr Neuropharmacol* 17(2): 108–28.

Inserra, A. 2018. Hypothesis: The psychedelic ayahuasca heals traumatic memories via a sigma 1 receptor-mediated epigenetic-mnemonic process. *Front Pharmacol* 9: 330.

Kuypers, K.P.C. 2019. Psychedelic medicine: The biology underlying the persisting psychedelic effects. *Med Hypotheses* 125: 21–4.

Lansky, E.P., S. Lansky, and H.M. Paavilainen. 2017. *Harmal: The Genus Peganum*. Boca Raton, FL: CRC Press.

Lansky, E.S. 2020a. [Remembering the self: A review of Dr Joseph Tafur's *The Fellowship of the River: A Medical Doctor's Exploration into Traditional Amazonian Plant Medicine*.] [In Hebrew.] *Harefuah* 159(1): 55–7.

Lansky, E.S. 2020b. Novel harmino-ocudelic tuning (HOT) for ocular disorders. *Med Hypotheses* 143: 109834.

Liu, C.H., W.L. Chu, S.C. Liao, C.C. Yang, and C.C. Lin. 2019. Syrian rue seeds interacted with acacia tree bark in an herbal stew resulted in N,N-dimethyltryptamine poisoning. *Clin Toxicol (Phila)* 57(10): 867–9.

Mate, G. 2017. Foreword to *The Fellowship of the River: A Medical Doctor's Exploration into Traditional Amazonian Plant Medicine*, by J. Tafur. Phoenix, AZ: Espiritu Books.

Maurice, T., and T.P. Su. 2009. The pharmacology of sigma-1 receptors. *Pharmacol Ther* 124(2): 195–206.

Nardai, S., M. László, A. Szabó, et al. 2020. N,N-dimethyltryptamine reduces infarct size and improves functional recovery following transient focal brain ischemia in rats. *Exp Neurol* 327: 113245.

Newman, P.D. 2017. *Alchemically Stoned: The Psychedelic Secret of Freemasonry*. Myrtle, MS: Laudable Pursuit Press.

Ott, J. 1994. *Ayahuasca Analogues: Pangaean Entheogens*. Kennewick, WA: Natural Products Co.

Pawar, R.S., E. Grundel, A.R. Fardin-Kia, and J.I. Rader. 2014. Determination of selected biogenic amines in *Acacia rigidula* plant materials and dietary supplements using LC-MS/MS methods. *J Pharm Biomed Anal* 88: 457–66.

Rickli, A., M.C. Hoener, and M.E. Liechti. 2019. Pharmacological profiles of compounds in preworkout supplements ("boosters"). *Eur J Pharmacol* 859: 172515.

Sanches, R.F., F. de Lima Osório, R.G. Dos Santos, et al. 2016. Antidepressant effects of a single dose of ayahuasca in patients with recurrent depression: A SPECT study. *J Clin Psychopharmacol* 36(1): 77–81.

Schenberg, E.E. 2013. Ayahuasca and cancer treatment. *SAGE Open Med* 1: 2050312113508389.

Strassman, R. 2001. *The Spirit Molecule: A Doctor's Revolutionary Research into the Biology of Near-Death and Mystical Experiences*. Rochester, VT: Park Street Press.

Szabo, A. 2015. Psychedelics and immunomodulation: Novel approaches and therapeutic opportunities. *Front Immunol* 6: 358.

Tafur, J. 2017. *The Fellowship of the River: A Medical Doctor's Exploration into Traditional Amazonian Plant Medicine*. Phoenix, AZ: Espiritu Books.

van Waarde, A., A.A. Rybczynska, N.K. Ramakrishnan, K. Ishiwata, P.H. Elsinga, and R.A. Dierckx. 2015. Potential applications for sigma receptor ligands in cancer diagnosis and therapy. *Biochim Biophys Acta* 1848(10 Pt B): 2703–14.

Yun, J., K. Kwon, J. Choi, and C.H. Jo. 2017. Monitoring of the amphetamine-like substances in dietary supplements by LC-PDA and LC-MS/MS. *Food Sci Biotechnol* 26(5): 1185–90.

7 Acacias and Metabolic Syndrome

Metabolic syndrome is a complex group of symptoms with, putatively, a common etiology. Classically, the syndrome (also known as Syndrome X) was first recognized over 70 years ago, and formally, it was recognized about 30 years ago (Balkau et al. 2007). It may be defined as a "constellation of signs and symptoms that increase a patient's risk of developing heart disease and diabetes mellitus" (Kassi et al. 2011) which include central obesity, hypertension, hyperglycemia, and atherogenic dyslipidemia (Hoffman et al. 2015). Insulin resistance (Huang 2009) has been added to this list as a major presumptive "source of pathogenesis" (Shin et al. 2013).

Vitamin D deficiency may be increasingly added to the list of sources of pathogenesis as well, as it adversely affects the cardiovascular system, increases insulin resistance and obesity, and stimulates the renin-angiotensin-aldosterone system that causes hypertension. Prasad and Kochar (2016) found that the vitamin D receptor is almost always present in immune, vascular, myocardial, and pancreatic beta cells.

Ultimately, however, insulin resistance and the symptoms that descend from it can all be considered to be part of a long-term chronic inflammatory disturbance, a disease of inflammation. Indeed, elevated circulating levels of C-reactive protein, fibrinogen, serum amyloid A, and pro-inflammatory cytokines and chemokines are among its features, generated primarily from adipose tissue, or fat, notoriously from belly fat. Metabolic syndrome affects 35% of Americans directly, and more and more kids and teens are getting it (Pergher et al. 2010; Reddy et al. 2019; Tagi et al. 2020; Wittcopp and Conroy 2016).

Although diagnosis guidelines for children and adolescents still require extrapolating, according to Swarup et al. (2020), the current general rules for adults, which are continually updated, are as follows:

- Waist circumference > 40 in. in men, > 35 in. in women
- Triglycerides =/> 150 milligrams per deciliter of blood (mg/dL)
- Decreased high-density lipoprotein cholesterol (HDL) =/< 40 mg/dL in men or =/< 50 mg/dL in women
- Fasting glucose =/> 100 mg/dL
- Blood pressure: systolic =/> 130 mmHg and/or diastolic =/> 85 mmHg

In the widest sense, metabolic syndrome, with chronic inflammation and insulin resistance at its core, may in fact underlie or at least exacerbate many illnesses beyond cardiovascular disease and diabetes, including, for example, schizophrenia (Bora et al. 2018), with their usual treatment also contributing to metabolic syndrome (Ijaz et al. 2018), autism (Rivell and Mattson 2019), psoriasis (Singh et al. 2016), inflammatory skin conditions (Steele et al. 2019), polycystic ovary syndrome (Ali 2015), atopic dermatitis (Ali et al. 2018), osteoarthritis (Gao et al. 2020), sarcopenic obesity (Khadra et al. 2020), stroke (Roever et al. 2018), urological disease (Sáenz Medina and Carballido Rodríguez 2016), gout (Thottam et al. 2017), clotting disorders (Mosimah et al. 2019), and endometrial cancer (Yang and Wang 2019).

In addition to afflicting 25–35% of the population of the United States, metabolic syndrome also affects, on average, about 25% of the inhabitants of the Middle East (Ansarimoghaddam et al. 2018)

DOI: 10.1201/9780429440946-7

including Iran (Mazloomzadeh et al. 2018), with similar percentages in other parts of the world (Cameron et al. 2004). In Taiwan, Yen et al. (2008) found a link between chewing betel nuts and metabolic syndrome along with other traditional risk factors like "being a couch potato," overeating, and smoking tobacco.

Sarcopenic obesity, i.e., obesity in the presence of loss of muscle mass, is a not uncommon associated feature of aging and a common comorbidity of, if not a risk factor for, obesity. The obesity associated with metabolic syndrome is frequently of this type, as is ectopic obesity, which means fat in the wrong places, and specifically abdominal obesity, which is a metric for the syndrome, as noted earlier. In short, aging predisposes to metabolic syndrome, as does ectopic fat, which also fosters insulin resistance. The symptoms begin to become the causes, and the entire process, complete with craving for sugar, eating the sugar, which causes the syndrome, and more craving for sugar because of the syndrome, can continue *ad infinitum* or until a dramatic health event or some mysterious comet breaks the cycle. Fructose, especially in soft drinks, appears to be one of the most serious public health risks associated with metabolic syndrome (Narain et al. 2017; Taskinen et al. 2019). "[M]agnesium metabolism alterations, systemic and hypothalamic inflammation, shortening of telomere length, epigenetics, and circadian rhythm disturbances" may all be part of unhealthy, i.e., metabolic syndrome-prone, aging (Dominguez and Barbagallo 2016). Further, persons with metabolic syndrome are more at risk for infectious diseases, as indicated by the COVID euphemism, "pre-existing conditions." Metabolic syndrome predisposes to COVID-19 and its severe outcome, and, reciprocally, COVID-19 damages heart (Li, B. et al. 2020). Most of these pre-existing conditions are clearly related to the metabolic syndrome and its advanced manifestations related to glucose metabolism and cardiovascular illness. Metabolic syndrome may also encompass neoplasia, which, like schizophrenia, is a highly developed manifestation of the metabolic syndrome, though its classic manifestations are diabetes and cardiovascular illness, including hypertension and dyslipidemia.

Public health measures include educating the public against sedentary lifestyles and sugar consumption, as well as, according to one position, aggressively vaccinating the older population, who, because of their predisposition to metabolic syndrome, are more susceptible to vaccine-preventable disease (Frasca and Blomberg 2020).

One direction for prevention and even amelioration of metabolic syndrome is the so-called "Mediterranean diet," based on foods and especially plant products of Mediterranean origin (Bagetta et al. 2020). The emphasis on the therapeutic quality of such a diet includes appreciation for the role of polyphenols, most especially polyphenols of plant origin, including those in dry red wine, olive oil, olives, saffron, licorice, garlic (Hosseini and Hosseinzadeh 2015), hot peppers (Sanati et al. 2018), onions, *Citrus* fruits, almonds, pistachios, rosemary, oregano, and capers (Lansky et al. 2013), especially in maintaining and improving health. The diet impacts cardiovascular health, related to metabolic syndrome. As well, the foods in the Mediterranean Diet for metabolic syndrome prevention and amelioration collectively should and do maintain a low glycemic index (Jones et al. 2012), meaning that they release more slowly their sugars into the blood, which causes a slow rather than a rapid increase in blood sugar.

Such a diet may be advantageous by diminishing, controlling, or preventing metabolic syndrome, especially when capped with coffee (Baspinar et al. 2017) and/or tea (Li, X. et al. 2020).

In addition to coffee and tea, a number of herbs have been found to be advantageous for treating or preventing metabolic syndrome. Among these are *Crataegus pinnatifida* (Dehghani et al. 2019), *Ginkgo biloba* (Eisvand et al. 2020), non-psychoactive cannabinoids (Silvestri et al. 2015), *Silybum marianum* (Tajmohammadi et al. 2018), *Garcinia mangostana* (Tousian Shandiz et al. 2017), probiotics (Tenorio-Jiménez et al. 2020), *Phaseolus vulgaris*, *Garcinia cambogia*, *Nigella sativa*, Pu'er tea, *Irvingia gabonensis*, and *Caralluma fimbriata* (Payab et al. 2020). *Acacia s.l.* emerges from this melee for its seminal contribution to the fray.

Gum arabic, as is customary, is the first acacia product to be tried. Since this product is essentially no more than water-soluble fibers, its effect is almost more nutraceutical than pharmacological. In fact, giving gum acacia (and pectin) to people with metabolic syndrome to eat on a regular basis improved fasting endogenous glucose turnover, but it didn't change insulin resistance in the periphery or inflammatory lipolysis (Pouteau et al. 2010).

TNF-α-stimulated lipolysis *was*, however, along with peripheral insulin resistance, inhibited by proanthocyanidins from *Vachellia nilotica* (supplemented by an extract of flowers from *Humulus lupulus*), which also improved glycemic control in mice (Tripp et al. 2012). In a 12-week clinical trial, a control arm ate a modified Mediterranean diet, and the experimental group received the same diet with the supplement containing *V. nilotica* and hops. The experimental groups showed less ($p < 0.05$) total cholesterol, LDL-C, non-high-density lipoprotein cholesterol (non-HDL-C), cholesterol/HDL-C, triglyceride/HDL-C, apolipoprotein (apo) B, apo B/apo A-1, and homocysteine, a smaller total LDL particle number, and a greater HDL particle number (Lerman et al. 2008, 2010). Additional studies with the combined extract consisting of five parts *V. nilotica* and 1 part *H. lupulus* synergistically increased triglyceride content and adiponectin secretion in 3T3-L1 adipocytes in an *in vitro* hyperinsulinemic simulation and reduced glucose or insulin in obese mice.

In the second clinical trial, tablets containing either the proprietary combination, 100 mg RIAA (the hops compounds) and 500 mg of *V. nilotica* extract, or a placebo were administered to subjects with metabolic syndrome (3 tablets/day, n = 35; 6 tablets/day, n = 34; or a placebo, n = 35) for 12 weeks. The subjects taking the extract exhibited greater reductions in serum triglycerides, triglycerides: high-density lipoproteins, and fasting insulin relative to controls (Minich et al. 2010).

The careful and thorough work of Sugiyama-san and his colleagues (Ikarashi et al. 2011) from the Department of Clinical Pharmacokinetics at Tokyo's Hoshi University is an inspiration for its comprehensiveness and its excellence, directed to the development of a complex phytopharmaceutical from the bark of *Acacia mearnsii* specifically targeted to the prevention and treatment of metabolic syndrome. Their *in vivo* study of the Australian "black wattle" in a laboratory in Japan is the cutting edge of true and pure pharmacognostical development in the world today.

(a) (b)

FIGURE 7.1 *Acacia mearnsii*, black wattle. (a) *A. mearnsii* bark. (9/11/2011, Dargo, Victoria, Australia: by John Tann, CC BY 2.0, via Flickr, www.flickr.com/photos/31031835@N08/6285994716.) (b) *A. mearnsii* trees in a forest. (4/26/2017, View trail, Hawea Pl Olinda, Maui, Hawaii: by Forest & Kim Starr, CC BY 4.0, via Starr Environmental, www.starrenvironmental.com/images/image/?q=35061389272.)

There are several salient reasons for this accolade, which may sound hyperbolic at first glance. These reasons may be summarized as follows:

1. The goal is clear from the abstract: To find a way to use the *extract* to treat *metabolic syndrome.*
2. Metabolic syndrome is an appropriate, timely, and meaningful target for pharmaceutical development. Its sufferers represent, estimated conservatively, one-fifth of the world's population, and the number is growing. That is quite a market! And there are few, if any, actual treatments for metabolic syndrome!
3. The authors unabashedly state their objective as the development of an *extract* of the *A. mearnsii* bark, not, like nearly all researchers in this field, paying lip service to the plant from which the extract came but actually thinking right away about either a patentable mixture with another herb or, alternatively, a pure chemical isolated from the plant to be developed synthetically into still more powerful pure chemicals. In this case, the complex extract itself, no matter what type of standardization, is and will be the medicine to be taken to market, and a mixture with other, typically inactive materials (i.e., adulteration) or ultra-purification to the point of single molecules (missing the power of synergy among plant components) will be eschewed. The simple complex extract itself (Lansky and Von Hoff 2005) is a worthy and strategically correct pharmaceutical objective. A complex extract is strategic precisely because it is complex. The different phytochemicals in their natural environments, or "natural phytochemical matrices," make it possible for many arrows to be shot at more than one target. This should not be seen as messy and hard to control (which it is!) but rather as a mimicry of Nature itself, which is nothing if not redundant! Diseases work through multi-targeting (of human receptors, etc.), so why shouldn't the medicines also be multi-targeted? This is not to say that single compound drugs are not also multi-targeted—they usually are, since they affect genes that are pleiotropic, and so even pure chemical drugs can exert multi-targeted effects. However, when building greater complexity into the system in order to more effectively multi-target, it is nigh impossible, even with the best biocomputers of today, to create complexes just by mixing together selected pure chemicals as a means of creating effective biocomplexity. One example, admittedly from quite a few years ago, was by the Sunkist chemists in Sarasota, Florida, looking to design an artificial orange extract that tasted exactly like oranges. Even though their knowledge of the compounds in oranges increased every few years, and their measuring devices became increasingly sophisticated, the newest synthetic formulas never tasted like oranges! Similarly, for designing complex drugs for the treatment of a condition as pleomorphic as metabolic syndrome, the "heavy lifting" can and should be done by the plants themselves, as they have their own innate, inherent, internally balanced, and genetically conserved complexity, which justifies the complexity of human diseases! That is why using complex extracts is a practical way to ensure both efficacy and safety. One example from Professor Andrew Weil, MD, concerned the drug digitalis, originally derived from the foxglove plant. The old textbooks described the three stages of toxicity: (1) Nausea and vomiting, (2) Arrhythmia, and (3) Death. Later, as the foxglove powder was replaced by a pure chemical, the stages of digitalis poisoning were reduced to two: (1) Arrhythmia and (2) Death.

Ikarashi's animal study (2011) used an extract of *A. mearnsii* bark in which were identified two "unique" catechin-like flavan-3-ols: *robinetinidol* and *fisetinidol*. The *A. mearnsii* extract was studied in KKAy diabetic and obese mice. The control mice were fed a normal scientific mouse diet for 7 weeks. A second group of mice was fed a high-fat diet, and a third group received a high-fat diet including the *A. mearnsii* bark extract. Results showed, as expected, a genetic baseline of glucose intolerance and obesity in all mice, with the high-fat group showing relative to the control group,

TABLE 7.1
Regulation of Selected Genes/Proteins in Response to Extract of *Acacia mearnsii* Stem Bark

Source	Gene/Protein	Regulation ↑ ↓
	PPARα	↑
Skeletal muscle	PPARδ	↑
mRNA expression of energy expenditure-related genes	CPT1	↑
	ACO	↑
	UCP3	↑
Liver	SREBP-1c	↓
Fat acid synthesis-related genes	ACC	↓
	FAS	↓
	Adiponectin	↑
White adipose tissue	TNF-α	↓

Source: After Ikarashi et al. (2011).

high serum glucose, body weight, and plasma insulin. These changes were reversed in the high-fat group of animals that also received the *A. mearnsii* extract. Further up- and downregulation of selected factors in response to the black wattle extract are summarized in Table 7.1.

As can be seen, research into the potential applications of *Acacia s.l.* directed at prevention and/or treatment of metabolic syndrome is still in an early stage. Potential avenues for research and development in this diverse area suggest new ideas and hypotheses. One unexplored area is the mechanism or mechanisms involved in mind-body regulation in general and the role of psychedelics catalyzing potentially long-lasting changes in lifestyle and putative "somatodelic" mechanisms (Lansky et al. 2017).

One phenomenon regarding psychedelic substances is of further interest. Many people who have had psychedelic experiences with substances like Ayahuasca or potentially similar experiences with DMT-bearing parts of wattles turn inward after the experience, often to an extent not previously thought likely or possible, with many beginning yoga and meditation practices (Millière et al. 2018; Richert and DeCloedt 2018). These practices are mostly a natural inclination following psychedelic experiences, and help to re-summon the experience on a daily basis albeit in an attenuated form. Such practices are directly beneficial for the prevention and amelioration of metabolic syndrome (Cramer et al. 2016; Levine et al. 2017; Tyagi et al. 2016; Vaccarino et al. 2013). On the vanguard is the idea that human physiology influences the physiology of plants and *vice versa*. Professor Anna Arola-Arnal and her colleagues in Tarragona, Spain, might say that the balance of polyphenols from seasonal fruits and vegetables affects the cycles of the human consumer (Arola-Arnal et al. 2019).

REFERENCES

Ali, A.T. 2015. Polycystic ovary syndrome and metabolic syndrome. *Ceska Gynekol* 80(4): 279–89.

Ali, Z., C.S. Ulrik, T. Agner, and S.F. Thomsen. 2018. Association between atopic dermatitis and the metabolic syndrome: A systematic review. *Dermatology* 234(3–4): 79–85.

Ansarimoghaddam, A., H.A. Adineh, I. Zareban, S. Iranpour, A. HosseinZadeh, and Kh. Framanfarma. 2018. Prevalence of metabolic syndrome in Middle East countries: Meta-analysis of cross-sectional studies. *Diabetes Metab Syndr* 12(2): 195–201.

Arola-Arnal, A., Á. Cruz-Carrión, C. Torres-Fuentes, et al. 2019. Chrononutrition and polyphenols: Roles and diseases. *Nutrients* 11(11): 2602.

Bagetta, D., A. Maruca, A. Lupia, et al. 2020. Mediterranean products as promising source of multi-target agents in the treatment of metabolic syndrome. *Eur J Med Chem* 186: 111903.
Balkau, B., P. Valensi, E. Eschwège, and G. Slama. 2007. A review of the metabolic syndrome. *Diabetes Metab* 33(6): 405–13.
Baspinar, B., G. Eskici, and A.O. Ozcelik. 2017. How coffee affects metabolic syndrome and its components. *Food Funct* 8(6): 2089–101.
Bora, E., B.B. Akdede, and K. Alptekin. 2017. The relationship between cognitive impairment in schizophrenia and metabolic syndrome: A systematic review and meta-analysis [correction published in 2018 in *Psychol Med* 48(7): 1224]. *Psychol Med* 47(6): 1030–40.
Cameron, A.J., J.E. Shaw, and P.Z. Zimmet. 2004. The metabolic syndrome: Prevalence in worldwide populations. *Endocrinol Metab Clin North Am* 33(2): 351–75.
Cramer, H., J. Langhorst, G. Dobos, and R. Lauche. 2016. Yoga for metabolic syndrome: A systematic review and meta-analysis. *Eur J Prev Cardiol* 23(18): 1982–93.
Dehghani, S., S. Mehri, and H. Hosseinzadeh. 2019. The effects of *Crataegus pinnatifida* (Chinese hawthorn) on metabolic syndrome: A review. *Iran J Basic Med Sci* 22(5): 460–8.
Dominguez, L.J., and M. Barbagallo. 2016. The biology of the metabolic syndrome and aging. *Curr Opin Clin Nutr Metab Care* 19(1): 5–11.
Eisvand, F., B.M. Razavi, and H. Hosseinzadeh. 2020. The effects of *Ginkgo biloba* on metabolic syndrome: A review. *Phytother Res* 34(8): 1798–1811.
Frasca, D., and B.B. Blomberg. 2020. The impact of obesity and metabolic syndrome on vaccination success. *Interdiscip Top Gerontol Geriatr* 43: 86–97.
Gao, Y.H., C.W. Zhao, B. Liu, et al. 2020. An update on the association between metabolic syndrome and osteoarthritis and on the potential role of leptin in osteoarthritis. *Cytokine* 129: 155043.
Hoffman, E.L., T. VonWald, and K. Hansen. 2015. The metabolic syndrome. *S D Med* (special edition) 2015: 24–8.
Hosseini, A., and H. Hosseinzadeh. 2015. A review on the effects of *Allium sativum* (Garlic) in metabolic syndrome. *J Endocrinol Invest* 38(11): 1147–57.
Huang, P.L. 2009. A comprehensive definition for metabolic syndrome. *Dis Model Mech* 2(5–6): 231–7.
Ijaz, S., B. Bolea, S. Davies, et al. 2018. Antipsychotic polypharmacy and metabolic syndrome in schizophrenia: A review of systematic reviews. *BMC Psychiatry* 18(1): 275.
Ikarashi, N., T. Toda, T. Okaniwa, K. Ito, W. Ochiai, and K. Sugiyama. 2011. Anti-obesity and anti-diabetic effects of acacia polyphenol in obese diabetic KKAy mice fed high-fat diet. *Evid Based Complement Alternat Med* 2011: 952031.
Jones, J.L., M. Comperatore, J. Barona, et al. 2012. A Mediterranean-style, low-glycemic-load diet decreases atherogenic lipoproteins and reduces lipoprotein (a) and oxidized low-density lipoprotein in women with metabolic syndrome. *Metabolism* 61(3): 366–72.
Kassi, E., P. Pervanidou, G. Kaltsas, and G. Chrousos. 2011. Metabolic syndrome: Definitions and controversies. *BMC Med* 9: 48.
Khadra, D., L. Itani, Y. Chebaro, et al. 2020. Association between sarcopenic obesity and metabolic syndrome in adults: A systematic review and meta-analysis. *Curr Cardiol Rev* 16(2): 153–62.
Lansky, E.P., S. Lansky, and H.M. Paavilainen. 2017. *Harmal: The Genus Peganum.* Boca Raton, FL: CRC Press.
Lansky, E.P., H.M. Paavilainen, and S. Lansky. 2013. *Capers: The Genus Capparis.* Boca Raton, FL: CRC Press.
Lansky, E.P., and D.D. Von Hoff. 2005. Complex and simple. *Leuk Res* 29(6): 601–2.
Lerman, R.H., D.M. Minich, G. Darland, et al. 2008. Enhancement of a modified Mediterranean-style, low glycemic load diet with specific phytochemicals improves cardiometabolic risk factors in subjects with metabolic syndrome and hypercholesterolemia in a randomized trial. *Nutr Metab (Lond)* 5: 29.
Lerman, R.H., D.M. Minich, G. Darland, et al. 2010. Subjects with elevated LDL cholesterol and metabolic syndrome benefit from supplementation with soy protein, phytosterols, hops rho iso-alpha acids, and *Acacia nilotica* proanthocyanidins. *J Clin Lipidol* 4(1): 59–68.
Levine, G.N., R.A. Lange, C.N. Bairey-Merz, et al. 2017. Meditation and cardiovascular risk reduction: A scientific statement from the American Heart Association. *J Am Heart Assoc* 6(10): e002218.
Li, B., J. Yang, F. Zhao, et al. 2020. Prevalence and impact of cardiovascular metabolic diseases on COVID-19 in China. *Clin Res Cardiol* 109(5): 531–8.
Li, X., W. Wang, L. Hou, et al. 2020. Does tea extract supplementation benefit metabolic syndrome and obesity? A systematic review and meta-analysis. *Clin Nutr* 39(4): 1049–58.
Mazloomzadeh, S., Z. Rashidi Khazaghi, and N. Mousavinasab. 2018. The prevalence of metabolic syndrome in Iran: A systematic review and meta-analysis. *Iran J Public Health* 47(4): 473–80.

Millière, R., R.L. Carhart-Harris, L. Roseman, F.M. Trautwein, and A. Berkovich-Ohana. 2018. Psychedelics, meditation, and self-consciousness. *Front Psychol* 9: 1475.

Minich, D.M., R.H. Lerman, G. Darland, et al. 2010. Hop and acacia phytochemicals decreased lipotoxicity in 3T3-L1 adipocytes, db/db mice, and individuals with metabolic syndrome. *J Nutr Metab* 2010: 467316.

Mosimah, C.I., P.J. Murray, and J.W. Simpkins. 2019. Not all clots are created equal: A review of deficient thrombolysis with tissue plasminogen activator (tPA) in patients with metabolic syndrome. *Int J Neurosci* 129(6): 612–18.

Narain, A., C.S. Kwok, and M.A. Mamas. 2017. Soft drink intake and the risk of metabolic syndrome: A systematic review and meta-analysis. *Int J Clin Pract* 71(2): 10.

Payab, M., S. Hasani-Ranjbar, N. Shahbal, et al. 2020. Effect of the herbal medicines in obesity and metabolic syndrome: A systematic review and meta-analysis of clinical trials. *Phytother Res* 34(3): 526–45.

Pergher, R.N.Q., M.E. de Melo, A. Halpern, M.C. Mancini, and Liga de Obesidade Infantil. 2010. Is a diagnosis of metabolic syndrome applicable to children? *J Pediatr (Rio J)* 86(2): 101–8.

Pouteau, E., V. Ferchaud-Roucher, Y. Zair, et al. 2010. Acetogenic fibers reduce fasting glucose turnover but not peripheral insulin resistance in metabolic syndrome patients. *Clin Nutr* 29(6): 801–7.

Prasad, P., and A. Kochhar. 2016. Interplay of vitamin D and metabolic syndrome: A review. *Diabetes Metab Syndr* 10(2): 105–12.

Reddy, P., D. Lent-Schochet, N. Ramakrishnan, M. McLaughlin, and I. Jialal. 2019. Metabolic syndrome is an inflammatory disorder: A conspiracy between adipose tissue and phagocytes. *Clin Chim Acta* 496: 35–44.

Richert, L., and M. DeCloedt. 2018. Supple bodies, healthy minds: Yoga, psychedelics and American mental health. *Med Humanit* 44(3): 193–200.

Rivell, A., and M.P. Mattson. 2019. Intergenerational metabolic syndrome and neuronal network hyperexcitability in autism. *Trends Neurosci* 42(10): 709–26.

Roever, L., E.S. Resende, A.L.D. Diniz, et al. 2018. Metabolic syndrome and risk of stroke: Protocol for an update systematic review and meta-analysis. *Medicine (Baltimore)* 97(15): e9862.

Sáenz Medina, J., and J. Carballido Rodríguez. 2016. Review of the pathophysiological aspects involved in urological disease associated with metabolic syndrome. Aspectos fisiopatológicos implicados en la patología urologica asociada al síndrome metabólico. Revisión bibliográfica. *Actas Urol Esp* 40(5): 279–87.

Sanati, S., B.M. Razavi, and H. Hosseinzadeh. 2018. A review of the effects of *Capsicum annuum* L. and its constituent, capsaicin, in metabolic syndrome. *Iran J Basic Med Sci* 21(5): 439–48.

Shin, J.A., J.H. Lee, S.Y. Lim, et al. 2013. Metabolic syndrome as a predictor of type 2 diabetes, and its clinical interpretations and usefulness. *J Diabetes Investig* 4(4): 334–43.

Silvestri, C., D. Paris, A. Martella, et al. 2015. Two non-psychoactive cannabinoids reduce intracellular lipid levels and inhibit hepatosteatosis. *J Hepatol* 62(6): 1382–90.

Singh, S., P. Young, and A.W. Armstrong. 2016. Relationship between psoriasis and metabolic syndrome: A systematic review. *G Ital Dermatol Venereol* 151(6): 663–77.

Steele, C.E., D. Morrell, and M. Evans. 2019. Metabolic syndrome and inflammatory skin conditions. *Curr Opin Pediatr* 31(4): 515–22.

Swarup, S., A. Goyal, Y. Grigorova, and R. Zeltser. 2020. Metabolic syndrome. In: *StatPearls* [Internet]. Treasure Island, FL: StatPearls Publishing.

Tagi, V.M., S. Samvelyan, and F. Chiarelli. 2020. Metabolic syndrome in children. *Minerva Pediatr* 72(4): 312–25.

Tajmohammadi, A., B.M. Razavi, and H. Hosseinzadeh. 2018. *Silybum marianum* (milk thistle) and its main constituent, silymarin, as a potential therapeutic plant in metabolic syndrome: A review. *Phytother Res* 32(10): 1933–49.

Taskinen, M.R., C.J. Packard, and J. Borén. 2019. Dietary fructose and the metabolic syndrome. *Nutrients* 11(9): 1987.

Tenorio-Jiménez, C., M.J. Martínez-Ramírez, Á Gil, and C. Gómez-Llorente. 2020. Effects of probiotics on metabolic syndrome: A systematic review of randomized clinical trials. *Nutrients* 12(1): 124.

Thottam, G.E., S. Krasnokutsky, and M.H. Pillinger. 2017. Gout and metabolic syndrome: A tangled web. *Curr Rheumatol Rep* 19(10): 60.

Tousian Shandiz, H., B.M. Razavi, and H. Hosseinzadeh. 2017. Review of *Garcinia mangostana* and its xanthones in metabolic syndrome and related complications. *Phytother Res* 31(8): 1173–82.

Tripp, M.L., G. Darland, V.R. Konda, et al. 2012. Optimized mixture of hops rho iso-alpha acids-rich extract and acacia proanthocyanidins-rich extract reduces insulin resistance in 3T3-L1 adipocytes and improves glucose and insulin control in db/db mice. *Nutr Res Pract* 6(5): 405–13.

Tyagi, A., M. Cohen, J. Reece, S. Telles, and L. Jones. 2016. Heart rate variability, flow, mood and mental stress during yoga practices in yoga practitioners, non-yoga practitioners and people with metabolic syndrome. *Appl Psychophysiol Biofeedback* 41(4): 381–93.

Vaccarino, V., K.A. Kondwani, M.E. Kelley, et al. 2013. Effect of meditation on endothelial function in Black Americans with metabolic syndrome: A randomized trial. *Psychosom Med* 75(6): 591–9.

Wittcopp, C., and R. Conroy. 2016. Metabolic syndrome in children and adolescents. *Pediatr Rev* 37(5): 193–202.

Yang, X., and J. Wang. 2019. The role of metabolic syndrome in endometrial cancer: A review. *Front Oncol* 9: 744.

Yen, A.M., L.S. Chen, Y.H. Chiu, B.J. Boucher, and T.H. Chen. 2008. A prospective community-population-registry based cohort study of the association between betel-quid chewing and cardiovascular disease in men in Taiwan (KCIS no. 19). *Am J Clin Nutr* 87(1): 70–8.

8 Acacias and Obesity

Obesity is, as is well known, a serious health problem worldwide (Shettar et al. 2017), especially in the United States (Arroyo-Johnson and Mincey 2016), with ~10% of preschool children suffering from or at high risk for the disorder (de Onis et al. 2010). Persons with obesity are more likely to contract COVID-19 and to have worse outcomes from it than expected (Kassir 2020). As a physiological disorder or disease, as opposed to a volitional disorder of behavior, the shift came only in 1985 in the scientific community and only in 2013 by the American Medical Association, after the discoveries of leptin and congeners (Coulter et al. 2018). Complications of obesity are legion, from metabolic and cardiac problems to vertebral disc degeneration (Ruiz-Fernández et al. 2019) and compromise of male fertility (El Salam 2018). The adipokines leptin and adiponectin are secreted in both prepubescent boys and girls, and they induce an early onset of puberty in girls and boys with obesity by affecting the hypothalamic-pituitary-gonadal (HPG) axis, thus revealing a further connection between sex hormones and obesity (Nieuwenhuis et al. 2020).

Obesity may be derivative of metabolic syndrome (Chapter 7), and it is considered to devolve from inflammation run amok, particularly when clinical insulin resistance is comorbid (Arroyo-Jousse et al. 2020; Das 2001; Izaola et al. 2015; Stolarczyk 2017). The obesity generates more inflammation, and so the cycle is vicious with unrestrained positive feedback (Kwaifa et al. 2020). For example, mice lacking the gene for nuclear factor E2-related factor 2 (NRF2), the so-called master antioxidant promoter, are more prone to obesity than normal mice, revealing the importance of antioxidant mechanisms for maintaining health (Wu et al. 2020). Nonetheless, excessive prescribing of dietary antioxidants, whether synthetically produced, such as food preservatives sodium benzoate or sodium sulfite, or natural, such as curcumin, decreases release of the appetite-suppressing hormone, leptin, from mouse adipocytes. In humans, such suppression of leptin contributes to obesity, most alarmingly in childhood (Mangge et al. 2013, 2017). Resistance of brain tissue to leptin is a hallmark of obesity (Eikelis and Esler 2005), and a relationship between serotonin and insulin secretion has been appreciated (Lernmark 1971).

Dietary tryptophan from tryptophan-rich foods may ameliorate depression through endogenous conversion of tryptophan to serotonin (Shabbir et al. 2013), as may a closer serotonin precursor, 5-hydroxytryptophan (Xu et al. 2020), and the "Western diet" of copious saturated fat and sugars aggravates pro-inflammatory and pro-diabetes metabolic parameters in serotonin transporter deficient mice (Veniaminova et al. 2020a). The Western diet-fed mice were also less glucose-tolerant and more insulin-resistant, with impaired motor coordination and novel object exploration and recognition, increased helplessness, dyslipidemia, and a non-alcoholic steatohepatitis (NASH)-like syndrome: liver steatosis and increased liver injury markers with decreased *serotonin* transporter protein (SERT) expression (Veniaminova et al. 2020b).

Dietary risk factors for obesity have been studied, and some interventions have succeeded, such as lowering the protein content of infant formulas, which reduces the risk of obesity in later life (Patro-Gołąb et al. 2018). Pharmacotherapeutics for obesity fall into three different categories: (1) drugs that reduce food intake, (2) drugs that alter metabolism, and (3) drugs that increase thermogenesis. Drugs acting at serotonin or other monoamine receptors can reduce food intake, and drugs that increase thermogenesis, such as combinations of ephedrine and caffeine, pend approval (Bray 2000; Bray and Ryan 2014). Harmine, the common alkaloid especially prominent in *Peganum harmala* and *Banisteriopsis caapi*, is another natural product known to increase adipocyte thermogenesis that awaits approval as an anti-obesity therapy (Nie et al. 2016). A natural herbal drug, *khat*—fresh or extracted leaves of the African plant *Catha edulis*—may both alter metabolism via its sympathomimetic properties and decrease food intake (Alshagga et al. 2016; Aziz et al. 2010).

DOI: 10.1201/9780429440946-8

Resveratrol, the chalcone of red wine and peanuts, is another natural product whose value in obesity has been demonstrated in two meta-analyses of clinical trials (Akbari et al. 2020; Tabrizi et al. 2020). The mechanism of resveratrol's anti-obesity effect is a model for therapeutic polyphenolic natural products, with both antioxidant and anti-inflammatory aspects at the cellular level (Kim et al. 2019). Resveratrol's anti-obesity effect may augment those of *curcumin* and *quercetin* (Zhao et al. 2017). *Tetrahydrocannabivarin* (THCV), a non-psychoactive component of cannabis and a metabolite of *tetrahydrocannabinol* (THC), has also been touted for its anti-obesity properties and therapeutic potential (Greenway and Kirwan 2019), likely related to its long chain fatty acids (Zhuang et al. 2020).

From this general anti-obesity phytoceutical milieu emerges the scattered phytochemistry across the huge *Acacia s.l.* supergenus. In one remarkable study, three Thai plants, *Eurycoma longifolia*, *Tiliacora triandra*, and *Acacia concinna*, were considered the most favorable of 70 tested as putative anti-obesogenic drugs due to their consistent activity in all three assays employed, *viz.*

FIGURE 8.1 *Acacia concinna*, the Ayurvedic *shikakai*, is known for its bright, beautiful red flowers and rich soap-giving seeds that make a very basic and serviceable shampoo that has been safely and effectively employed as a hair cosmetic in India for centuries. The tree is home to the gorgeous common India South Asia butterfly, *Pantoporia hordonia*. (a) *A. concinna* leaves and flower, botanical drawing. (In F. M. Blanco, *Flora de Filipinas*, ed. 3, t, 375 [Manila; Estab. tip. de Plana y ca, 1877–83], via www.plantillustrations.org, public domain, www.plantillustrations.org/illustration.php?id_illustration=62377.) (b) *Senegalia rugata* (= *A. concinna*) branch with leaves and mature seed pods. (By Ram Kumar Banskar, via Shutterstock, www.shutterstock.com/image-photo/senegalia-rugata-commonly-known-india-shikakai-1926541265.) (c) *Senegalia rugata* (= *A. concinna*) branches with flowers and buds. (3/1/2014, IISC Bangalore, Gulmohar Marg, Mathikere, Bengaluru, Karnataka 560012, India: by Ajit Ampalakkad, CC0 1.0, via iNaturalist, www.inaturalist.org/photos/45616977.)

FIGURE 8.2 *Pantoporia hordonia* loves its *Acacia concinna*! Yum!—*Pantoporia hordonia*, Common Lascar, in open wing position. (10/13/2019, Arunachal Pradesh, India; by Rejoice Gassah, CC BY 4.0, via iNaturalist, www.inaturalist.org/photos/63811651.)

pancreatic lipase inhibition, lipolysis enhancement, and lipid accumulation reduction (Ruangaram and Kato 2020). The *A. concinna*, "shampoo tree," whose pod and seed extracts are so saponin-rich that they have been used for shampoo in India for centuries, has been characterized and found to contain triterpenoid prosapogenols *concinnoside A*, *B*, *C*, *D*, and *E*, known glycosides *acaciaside*, *julibrosides A1* and *A3*, *albiziasaponin C*, and their aglycone, *acacic acid lactone* (Abul Gafur et al. 1997). Immunological adjuvant activities comparable to those of *Quillaja saponaria*, but with less surface tension, were noted for the *A. concinna* saponins (Kukhetpitakwong et al. 2006). *Acacinic acid* was earlier found in the seeds of *Acacia intsia* and *A. concinna* (Farooq et al. 1962). Three saponin *kinmoonasides* (*A*, *B*, *C*) and a monoterpene, *4-O-[(2E)-6-hydroxyl-2-hydroxymethyl-6-methyl-2,7-octadienoyl]-D-quinovopyranose*, all cytotoxic to human HT-1080 fibrosarcoma cells *in vitro*, were also discovered in the *A. concinna* extract (Tezuka et al. 2000).

Targeting oxidative and inflammatory pathways, extracts of other acacias also offer promise for treating obesity. "Wattle tannin," extracted from the bark of *Acacia mearnsii* with hot water, contained distinctive flavan-3-ols *fisetinidol*, *robinetinidol*, *catechin*, and *gallocatechin*; four biflavanoids, trans compounds *fisetinidol-(4α-8)-catechin*, *robinetinidol-(4α-8)-catechin*, *robinetinidol-(4α-8)-gallocatechin*, and *fisetinidol-(4β-8)-catechin*; and two triflavanoids, *robinetinidol-(4α-8")-robinetinidol (4'α-6")-gallocatechin* and *robinetinidol-(4α-8")-robinetinidol (4'α-6")-catechin*; dimeric proanthocyanidins, *robinetinidol-(4α-8)-catechin*, *fisetinidol-(4β-8)-catechin*, and *robinetinidol-(4β-8)-catechin*; and *leucofisetinidin*, *leucorobinetinidin*, *quercetin*, *myricetin*, *butin*, *butein*, *robtein*, *fisetinidol*, *robinetinidol*, *catechin*, *gallocatechin*, *fustin*,

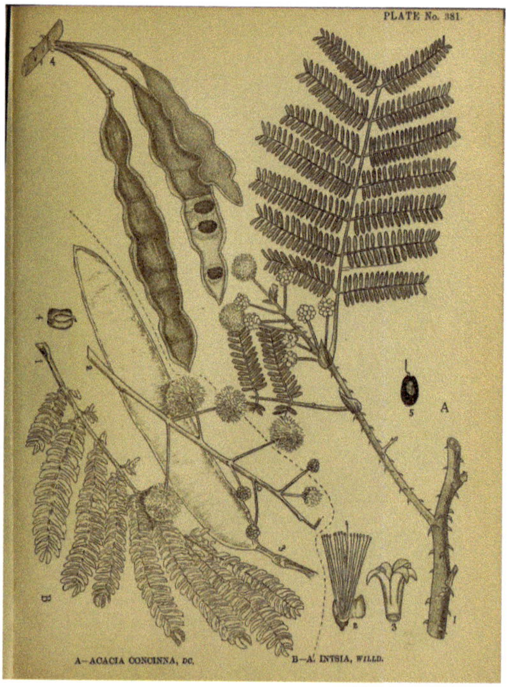

FIGURE 8.3 *Acacia intsia* is a perennial shrub or woody climber growing in subtropical Asian forests, dry grasslands, and shrubby regions. *A. intsia*, botanical drawing. (K. R. Kirtikar and B. D. Basu, *Indian medicinal plants, Plates* (1918) [Allahabad: Lalit Mohan Basu, 1918], via www.plantillustrations.org, public domain, www.plantillustrations.org/illustration.php?id_illustration=162042.)

dihydrorobinetin, *fisetin*, and *robinetin*, which caused *in vitro* inhibition of enzymes α-amylase, lipase, and glucosidase (Ogawa and Yazaki 2018). Examinations of the complex *in vivo* revealed anti-obesity and anti diabetes activities (Ikarashi et al. 2011, 2018).

An extract of the stem bark of *Vachellia nilotica* was the most efficient of a panel of over 200 natural products for reducing fasting as well as postprandial glucose and non-fasting insulin in a mouse model and decreasing TNF-α inhibition of adiponectin secretion by murine adipocytes *in vitro* (Babish et al. 2010). A medicinal food, prepared by sprouting *V. nilotica* seeds and pan frying them in oil, has been recommended as an ideal form for *per orum* administration (Vadivel and Biesalski 2012).

Other anti-obesity acacias of note include the Mexican species *Vachellia schaffneri* and *V. farnesiana*, extracts of whose seed pods elicited potent antioxidant effects *in vitro* in several established assays (Delgadillo Puga et al. 2015). These studies were started with the goal of finding possible new natural anti-obesity drugs.

Finally, the effects of gum arabic as an anti-obesity therapeutic have been documented. Gum arabic, the ubiquitous pharmaceutical additive, is always officially derived from the stem exudate of *Senegalia senegal*. In one clinical trial, it was found that a daily dose of 30 g/day gum arabic for 3 months significantly reduced basal metabolic index (BMI) and visceral adiposity index (VAI) ($p < 0.05$), while also effecting a significant decrease in the blood pressure of patients diagnosed with type 2 diabetes mellitus (Babiker et al. 2018). In an earlier, two-arm, double-blind clinical trial among 120 healthy females, the same regimen of 30 g/day for 6 weeks of gum arabic, a fiber that is indigestible by humans or animals, caused significant reductions in BMI by 0.32 (95% CI: 0.17 to 0.47; $p < 0.0001$) and body fat percentage by 2.18% (95% CI: 1.54 to 2.83; $p < 0.0001$), with side

FIGURE 8.4 Gum arabic. Notice the shape, texture, and transparency. (Photo by Picture Partners, via Shutterstock, www.shutterstock.com/image-photo/pieces-gum-arabic-on-white-background-470512106.)

effects limited to the first week including unpleasant oral viscosity, early morning nausea, mild diarrhea, and abdominal bloating (Babiker et al. 2012).

In wild-type C57BL mice, gum arabic in drinking water (100 g/L) resulted in decreased abundance of intestinal Na(+)-coupled glucose transporter (SGLT1) in jejunal brush border membrane vesicles, an effect associated with inhibition of SGLT 1 expression and associated high glucose stimulated weight gain (Nasir et al. 2010, 2013). In a later study in the same mice, gum arabic caused a decrease in body weight, epididymal fat mass, adipocyte size, hyperlipidaemia, hyperglycaemia, hyperinsulinemia, and the expression of leptin, interleukin 6, and tumor necrosis factor alpha, while upregulating the expressions of adiponectin and glucose transporter 4 in the epididymal fat. Gum arabic upregulated genes related to beta-oxidation, viz., carnitine palmitoyltransferase 1, peroxisome proliferator-activated receptor gamma co-activator 1β, and peroxisome proliferator-activated receptor alpha in the murine liver, as well as expressions of zonula occludens 1 and fasting-induced adipose factor in the distal ileum and proglucagon in the colon. All of these results showed that gum arabic fibers increased the amount of energy burned, improved the health of the intestines, and decreased inflammation. They also led to safe and effective weight loss (Jangra et al. 2019; Jangra and Pothuraju 2020).

Gut microbiota help drive obesity; their manipulation with dietary interventions holds clinical promise against obesity (Saad et al. 2016), as do ketogenic low carbohydrate diets (Li, R.J. et al. 2020). Gut microbiota regulate glucose homeostasis via serotonin (Martin et al. 2017, 2019), a legitimate new pharmaceutical target for both obesity and diabetes mellitus (Oh et al. 2016).

In fact, serotonin has a lot to do with obesity, including, for example, the association between obesity and migraine headaches, in both of which serotonin receptors and serotonin itself are involved (Peterlin et al. 2010). There is likely a genetic basis for this sharing of circuitry, as in the genetic overlap between cardiovascular and mood disorders (Amare et al. 2017). Carbohydrate craving in the extreme leads to obesity, and this craving is orchestrated in a sense by serotonin in order to provide more starting materials for further serotonin synthesis and the alleviation of dysphoria (Wurtman and Wurtman 1995, 2018). Yet, without serotonin, there can be no obesity and no insulin resistance (Yabut et al. 2020).

In a review of 2011, the authors state that serotonergic modulators had been the mainstay of anti-obesity treatments for the 35 years that had preceded (Bello and Liang 2011). Currently, there are very few anti-obesity drugs that are approved, due to regulatory concerns with the side effects of the available conventional agents (Bessesen and Van Gaal 2018). There were previous claims about 5-hydroxytryptophan, a direct serotonin precursor available commercially as a nutritional supplement, ameliorating binge eating associated with obesity (Birdsall 1998). Further research has revealed a key mechanism of the serotonergic network relevant for obesity treatment which involves one specific serotonin receptor, 5-HT2 C R, whose targeted antagonism by a safe and approved medication was sought (Burke and Heisler 2015). Safety and efficacy issues relating to specific 5-HT2 C R agonists have been considered (Cabrerizo García et al. 2013) as well as the potential for 5-HT2 C R agonists to treat not only food addictions related to obesity but also addictions to tobacco, alcohol, and classical "substances," i.e., "street drugs" and prescription pharmaceuticals (Higgins and Fletcher 2015; Higgins et al. 2013). Further research into such agents may confront the pervasive psychic dimension of obesity more directly through the effects of potential new drugs for mood and impulsivity disorders (Higgins et al. 2017).

According to one group in Turkey, acupuncture was claimed to aid obesity by exciting the satiety center in the hypothalamic ventromedial nuclei and via stimulation of the auricular branch of the vagus, thus raising serotonin levels, increasing stomach tone, and suppressing appetite (Cabýoglu et al. 2006). In a large study of obese subjects, a key pathway to serotonin from tryptophan was altered, relative to the pathway in normal matched controls, underscoring the importance serotonin plays in the etiology and maintenance of obesity (Cussotto et al. 2020). Certainly, the role of serotonin in establishing satiety and curtailing further feeding behavior, especially *vis-à-vis* serotonin receptor 5-HT2C, has been known (Feijó Fde et al. 1992), with a specific 5-HT2C R drug recently pending approval in Australia (Hocking et al. 2017). The drug in question, Lorcaserin, a novel serotonin 5-HT2C selective agonist that was shown in three phase III trials to significantly reduce weight and cardiovascular risk factors including, diabetes mellitus (Hurt et al. 2014), was approved by the US FDA for treatment of obesity in 2012 (Khorassani et al. 2015). Drugs targeting the 5-HT2C R may attenuate not only obesity (Marston and Heisler 2009) but possibly also co-occurring epilepsy (Heisler et al. 1998), depression, anxiety, or psychosis (Jensen et al. 2010). In order to offset the toxicity of the most effective pharmaceutical weight loss agent of its time (i.e., fenfluramine), a more specific new agonist of a putative subtype of 5-HT2C R was advised (Miller 2005), based on the uneven yet promising history of 5-HT2C R targeting drugs (Palacios et al. 2017).

Conversely, activation of the 5-HT1 receptor was known to stimulate overeating (Garattini et al. 1988), while the following two other serotonin receptors have also been recognized for their roles in experimental appetite and satiety and as possible pharmaceutical targets for experimental anti-obesity therapeutics, namely, (1) serotonin(1B) (rodent)/serotonin(1Dbeta) (human) receptor (5-HT(1B/1Dbeta)R) and (2) serotonin(6) receptor (5-HT(6)R) (Garfield and Heisler 2009). A proposal for an antagonist of 5-HT 3 as a novel anti-obesity agent has been advanced (Kurhe and Mahesh 2015). Serotonin receptors 1A, 1B, 2C, 2B, and 6 were all considered relevant to anti-obesity pharmacology (Vickers and Dourish 2004; Voigt and Fink 2015).

Building on new findings establishing serotonin as fundamentally responsible for regulating energy homeostasis through hepatic lipid metabolism, new drugs are being developed to antagonize the 5HT2A receptor, also as a means for treating obesity. The prototypes have acted peripherally to decrease liver weight and lipid accumulation in diet-induced obesity mice (Kim et al. 2020). Peculiar genetic variations in a serotonin 2A receptor were associated with arterial hypertension in a clinical population (Choi et al. 2020), but the receptor status was not changed by intensive dieting, although significant weight loss did occur (Rahm et al. 2004). Serotonin receptor 4 may also play a role in obesity, as its expression was upregulated, along with hypophagia, in response to restraint stress in mice (Rebholz et al. 2018). While most of the aforementioned receptors can be considered to have a hypophagic effect in animals, 5-HT6 mediates anorexia (Sargent and Henderson 2011).

Nevertheless, in spite of limitations due to only modest results and unwanted side effects, pharmacotherapy for obesity, especially by targeting serotonin and/or other monoamine receptors, has been widely attempted (Yabut et al. 2019). One example of an anti-obesity drug that has been approved is a step beyond the selective serotonin reuptake inhibitor (SSRI), namely, a serotonin and norepinephrine uptake inhibitor (SNRI). The drug not only resulted in significant weight loss but also improved lipid parameters, blood glucose, and insulin, factors pertinent to chronic obesity and metabolic syndrome (Hainer et al. 2006). Ongoing development of anti-obesity drugs aimed at the serotonergic and other monoaminergic systems continues, with the key goal being increased weight loss without related central nervous system or other side effects (Halford et al. 2011; Simonyi et al. 2012). A review of various anti-obesity drugs in use in the United States, as well as their side effect profiles, was published (Joo and Lee 2014).

A high-fat diet is correlated with anti depressant effects *in vivo*, but if the animals have been genetically modified so that their brains are serotonin deficient, the anti depressant effects associated with the high-fat diet do not occur (Karth et al. 2019). Further, rats genetically modified to have brains with “high serotonin tone” show statistically significantly more obesity than rats engineered to have “low serotonin tone” (Kesić et al. 2020).

New anti-obesity drugs may target the adipocyte peptide hormone, leptin, which is also capable of triggering signals of satiety from the hypothalamus, in lieu of serotonin-based drugs. In both cases, co-treatment of both obesity and the comorbid major depression is indicated in many patients (Haleem 2016). New serotonin-targeting drugs may also aim to simultaneously influence non-serotonergic pathways and receptors within the satiety-regulating hypothalamus (Harrold and Halford 2006). Promotion of central serotonin synthesis for signalizing was cited as the mechanism by which both mice and the nematode, *Caenorhabditis elegans*, induce fat loss (Lin et al. 2019). The flavone *luteolin* stimulates such *C. elegans* central serotonin synthesis (Lin et al. 2020), while the Zn transporter protein, ZnT8, downregulates serotonin synthesis from serotonin-containing enterochromaffin cells in the stomach. If the transporter is eliminated in a ZnT8 knockout mouse, exacerbation of obesity results (Mao et al. 2019). Similarly, the melanocortin hormones, of which adrenocorticotropic-releasing hormone (ACTH) is an example, also modulate the stomach's serotonin release, thus providing another natural control of obesity (Marston et al. 2011).

Due to the multifaceted aspects of obesity, an interdisciplinary approach to its therapy has been initiated and advocated (Habicht et al. 2017), and multifaceted polypharmacy has been similarly attempted and advised (Proietto et al. 2000). A possible contribution to obesity by the antidepressant drugs used to treat the depression causing the obesity has also been emphasized (Lee et al. 2016), as has the role of serotonin in modulating feeding behavior (Magalhães et al. 2010). Implementing anti-obesity pharmacotherapy following lifestyle changes may be advantageous (Manning et al. 2014).

Of great interest in all this is the gradual transition from the traditional view of serotonin as a strong central anorectic in brain, which drove serotonin research for obesity for decades, to one that more closely follows serotonin in the periphery, especially in adipocytes (Namkung et al. 2015).

In spite of their considerable therapeutic promise and the joy of their discovery, narrowing focus to specific 5-HT receptors and their subtypes and then either upregulating or downregulating one such sub-sub-receptor types as a means of therapy has not borne very good results. For one thing, the degree of weight loss has been only modest, and, most importantly, the safety of the drugs has not been adequate, causing, for example, hypertrophy of heart valves. Accordingly, most of these agents that have been developed have passed regulatory hurdles and were subsequently removed from the market, leaving only a few representative drugs to hold the line. Otherwise, the business of anti-obesity drugs, particularly with respect to serotonin targeting, has fallen short of earlier expectations.

However, as noted earlier, psychedelic drugs, in terms of their ability to influence serotonin receptors, perhaps by a long-term putative "tuning" (Lansky 2020b; Lansky et al. 2017), represent a novel approach to "targeting the serotonergic system" for the treatment of a metabolic disorder (Wyler et al. 2017). Reports describe lasting benefit from single, or from a modiucum of doses when administered under optimized mind-set ("set") and physical environment plus context ("setting"). How this takes place is unknown, but the essence of the tuning hypothesis is that the psychedelic "re-sets" the full spectrum of serotonin receptors in the system so that a new, more stable, healthier, and more efficient configuration remains, based on multiple changes at multiple serotonin receptors simultaneously through modulating the receptor "tuning dials" in the receptors' affinity for their substrate, i.e., serotonin. An analogy for this is restarting the computer, a cure that solves many problems.

In the psyche, this effect, though sometimes euphemized as "hallucinogenic," "entheogenic," "psychotomimetic," etc., is still best described by the name impressed upon such drugs by the British psychiatrist Humphrey Osmond in 1957 at a meeting of the New York Academy of Sciences, namely, *psychedelic*, calling it "clear, euphonious, and uncluttered by other associations" (Marks 2021), which means literally, from the Greek, "mind manifesting," and connotes the idea of mental reset.

Taking this to new realms, how about applying the Greek suffix "-delic" to serotonergic networks, not just in the central nervous system (CNS), but throughout the periphery as well, in somatic as well as psychic realms? For this general idea of "-delic" phenomena in the *soma*, the term "*somatodelic*" was proposed, to indicate an upheaval on the somatic level, specifically in serotonin receptor-studded T lymphocytes, homologous to what happens in neurons, i.e., when the brain is psychedelically inebriated (Lansky et al. 2017). There is also an ocular serotonergic network, and for a "-delic"-like effect from a psychedelic substance that radically improves vision in the real world through a putative re-tuning of serotonin receptors in the eye, the term "*ocudelic*" was coined (Lansky 2020b).

In that serotonin receptors also occur in adipocytes, and because serotonin-active drugs have been extensively utilized, in spite of their side effects, for their ability to enhance weight loss, it seems similarly reasonable to consider the use of serotonin receptor–modifying psychedelic substances to elicit a parallel "-delic" effect in the complex lipomic universe that is the serotonin-affected human lipidome (Gruenberg 2020; O'Donnell et al. 2020). Such a response, if obtained, could be dubbed "lipidelic." The dimethyltryptamine-rich bark, or most preferably seed pods, of selected trees of *Acacia s.l.*, supported by a guiding woody "banister" like the reversible monoamine oxidase inhibitor harmine and its fellow harmala alkaloids harmaline, etc., found in the Vine of the Soul, *Banisteriopsis caapi*, or in the hypothesized "soma" of the ancients, *Peganum harmala* (Lansky et al. 2017), could be assessed for putative "lipidelic" activity, i.e., global and rapid lipidomic and cardiovascular amelioration, *in vitro*, *in vivo*, and in the clinic, under appropriate medical supervision and guidance by both music and therapists. A clinical manifestation of the grand target of such an in-depth "lipidelic" therapy catalyzed by a psychedelic compound or complex could herald a radical improvement of personal obesity and global health (Flanagan and Nichols 2018; Flanagan et al. 2019; Song 2005) while further underlining the physiological, serotonin-mediated, functional commonality between mind and belly, brain and gut (Agrawal 2020; Lyte et al. 2020; Pan et al. 2019; Salisbury et al. 2020; Szőke et al. 2020). The aforementioned harmine increases thermogenic energy expenditure in white adipose tissue, which delays the onset of diabetes caused by obesity (Sethi and Vidal-Puig 2007). It was also the only compound out of 3000 tested at the time that helped pancreatic beta cells grow back (Wang et al. 2015).

Finally, timing is of utmost importance. Even cells have their own biological clocks, a functionality that may be interrupted by serotonin. In one case, serotonin stopped brown adipocytes from differentiating *in vitro*. This shows that serotonin is linked to obesity (Rozenblit-Susan et al. 2019).

The common antidepressant fluoxetine was noted to have some weight loss potency in healthy overweight or obese adults, but the optimism was tempered by the propensity of such drugs to cause side effects, including dizziness, drowsiness, fatigue, insomnia, and nausea (Serralde-Zúñiga et al. 2019). Clearly, there is a need for non-toxic, natural anti-obesity treatments that can effectively restore the deficient serotonergic and dopaminergic signaling characterizing obesity (van Galen et al. 2018).

Pertinent further work by Charles Nichols's team with a synthetic, long-lasting psychedelic and selective 5-HT2a agonist, *(R)-2,5-dimethoxy-4-iodoamphetamine [(R)-DOI]*, led to improvements in biochemical parameters related to obesity in mice (Flanagan et al. 2019).

REFERENCES

Abul Gafur, M., T. Obata, F. Kiuchi, and Y. Tsuda. 1997. *Acacia concinna* saponins. I. Structures of prosapogenols, concinnosides A–F, isolated from the alkaline hydrolysate of the highly polar saponin fraction. *Chem Pharm Bull (Tokyo)* 45(4): 620–5.

Agrawal, L., M. Korkutata, S.K. Vimal, M.K. Yadav, S. Bhattacharyya, and T. Shiga. 2020. Therapeutic potential of serotonin 4 receptor for chronic depression and its associated comorbidity in the gut. *Neuropharmacology* 166: 107969.

Akbari, M., O.R. Tamtaji, K.B. Lankarani, et al. 2020. The effects of resveratrol on lipid profiles and liver enzymes in patients with metabolic syndrome and related disorders: A systematic review and meta-analysis of randomized controlled trials. *Lipids Health Dis* 19(1): 25.

Alshagga, M.A., M.A. Alshawsh, A. Seyedan, et al. 2016. Khat (*Catha edulis*) and obesity: A scoping review of animal and human studies. *Ann Nutr Metab* 69(3–4): 200–11.

Amare, A.T., K.O. Schubert, M. Klingler-Hoffmann, S. Cohen-Woods, and B.T. Baune. 2017. The genetic overlap between mood disorders and cardiometabolic diseases: A systematic review of genome wide and candidate gene studies. *Transl Psychiatry* 7(1): e1007.

Arroyo-Johnson, C., and K.D. Mincey. 2016. Obesity epidemiology worldwide. *Gastroenterol Clin North Am* 45(4): 571–9.

Arroyo-Jousse, V., A. Jaramillo, E. Castaño-Moreno, M. Lépez, K. Carrasco-Negüe, and P. Casanello. 2020. Adipokines underlie the early origins of obesity and associated metabolic comorbidities in the offspring of women with pregestational obesity. *Biochim Biophys Acta Mol Basis Dis* 1866(2): 165558.

Aziz, H.A., Y.T. Tan, K.K. Peh, and M.F. Yam. 2010. Direct effect of khat and garlic extracts on blood lipids contents: Preliminary *in vitro* study. *Obes Res Clin Pract* 4(4): e247–e342.

Babiker, R., K. Elmusharaf, M.B. Keogh, and A.M. Saeed. 2018. Effect of gum arabic (*Acacia Senegal*) supplementation on visceral adiposity index (VAI) and blood pressure in patients with type 2 diabetes mellitus as indicators of cardiovascular disease (CVD): A randomized and placebo-controlled clinical trial. *Lipids Health Dis* 17(1): 56.

Babiker, R., T.H. Merghani, K. Elmusharaf, R.M. Badi, F. Lang, and A.M. Saeed. 2012. Effects of gum Arabic ingestion on body mass index and body fat percentage in healthy adult females: Two-arm randomized, placebo controlled, double-blind trial. *Nutr J* 11: 111.

Babish, J.G., L.M. Pacioretty, J.S. Bland, D.M. Minich, J. Hu, and M.L. Tripp. 2010. Antidiabetic screening of commercial botanical products in 3T3-L1 adipocytes and db/db mice. *J Med Food* 13(3): 535–47.

Bello, N.T., and N.C. Liang. 2011. The use of serotonergic drugs to treat obesity—Is there any hope? *Drug Des Devel Ther* 5: 95–109.

Bessesen, D.H., and L.F. Van Gaal. 2018. Progress and challenges in anti-obesity pharmacotherapy. *Lancet Diabetes Endocrinol* 6(3): 237–48.

Birdsall, T.C. 1998. 5-Hydroxytryptophan: A clinically effective serotonin precursor. *Altern Med Rev* 3(4): 271–80.

Bray, G.A. 2000. A concise review on the therapeutics of obesity. *Nutrition* 16(10): 953–60.

Bray, G.A., and D.H. Ryan. 2014. Update on obesity pharmacotherapy. *Ann N Y Acad Sci* 1311: 1–13.

Burke, L.K., and L.K. Heisler. 2015. 5-Hydroxytryptamine medications for the treatment of obesity. *J Neuroendocrinol* 27(6): 389–98.

Cabrerizo García, L., A. Ramos-Leví, C. Moreno Lopera, and M.A. Rubio Herrera. 2013. Update on pharmacology of obesity: Benefits and risks. *Nutr Hosp* 28(Suppl 5): 121–7.

Cabýoglu, M.T., N. Ergene, and U. Tan. 2006. The treatment of obesity by acupuncture. *Int J Neurosci* 116(2): 165–75.

Choi, J.R., M. Jeon, and S.B. Koh. 2020. Association between serotonin 2A receptor (HTR2A) genetic variations and risk of hypertension in a community-based cohort study. *BMC Med Genet* 21(1): 5.

Coulter, A.A., C.J. Rebello, and F.L. Greenway. 2018. Centrally acting agents for obesity: Past, present, and future. *Drugs* 78(11): 1113–32.

Cussotto, S., I. Delgado, A. Anesi, et al. 2020. Tryptophan metabolic pathways are altered in obesity and are associated with systemic inflammation. *Front Immunol* 11: 557.

Das, U.N. 2001. Is obesity an inflammatory condition? *Nutrition* 17(11–12): 953–66.

de Onis, M., M. Blössner, and E. Borghi. 2010. Global prevalence and trends of overweight and obesity among preschool children. *Am J Clin Nutr* 92(5): 1257–64.

Delgadillo Puga, C., M. Cuchillo Hilario, J.G. Espinosa Mendoza, et al. 2015. Antioxidant activity and protection against oxidative-induced damage of *Acacia shaffneri* and *Acacia farnesiana* pods extracts: In vitro and in vivo assays. *BMC Complement Altern Med* 15(1): 435.

Eikelis, N., and M. Esler. 2005. The neurobiology of human obesity. *Exp Physiol* 90(5): 673–82.

El Salam, M.A.A. 2018. Obesity, an enemy of male fertility: A mini review. *Oman Med J* 33(1): 3–6.

Farooq, M.O., I.P. Varshney, and Z. Naim. 1962. [Saponins and Saponogenins. X. On the occurrence of acacinic acid in the seeds of *Acacia Concinna* DC. and *Acacia Intsia* Willd]. *Arch Pharm Ber Dtsch Pharm Ges* 295(67): 12–14.

Feijó Fde, M., M.C. Bertoluci, and C. Reis. 2011. Serotonina e controle hipotalâmico da fome: uma revisão [Serotonin and hypothalamic control of hunger: A review]. *Rev Assoc Med Bras (1992)* 57(1): 74–7.

Flanagan, T.W., and C.D. Nichols. 2018. Psychedelics as anti-inflammatory agents. *Int Rev Psychiatry* 30(4): 363–75.

Flanagan, T.W., M.N. Sebastian, D.M. Battaglia, T.P. Foster, E.L. Maillet, and C.D. Nichols. 2019. Activation of 5-HT2 receptors reduces inflammation in vascular tissue and cholesterol levels in high-fat diet-fed apolipoprotein E knockout mice. *Sci Rep* 9(1): 13444.

Garattini, S., A. Bizzi, S. Caccia, T. Mennini, and R. Samanin. 1988. Progress in assessing the role of serotonin in the control of food intake. *Clin Neuropharmacol* 11(Suppl 1): S8–S32.

Garfield, A.S., and L.K. Heisler. 2009. Pharmacological targeting of the serotonergic system for the treatment of obesity. *J Physiol* 587(1): 49–60.

Greenway, F.L., and J.P. Kirwan. 2019. Medical marijuana—an obesity problem or opportunity? *Int J Obes (Lond)* 43(4): 761–2.

Gruenberg, J. 2020. Life in the lumen: The multivesicular endosome. *Traffic* 21(1): 76–93.

Habicht, F., M.L. Guillermin-Spahr, and E. Bobbioni-Harsch. 2017. L'interdisciplinarité pour un traitement global du patient obèse [Interdisciplinarity for obese patient global management]. *Rev Med Suisse* 13(554): 623–5.

Hainer, V., K. Kabrnova, B. Aldhoon, M. Kunesova, and M. Wagenknecht. 2006. Serotonin and norepinephrine reuptake inhibition and eating behavior. *Ann N Y Acad Sci* 1083: 252–69.

Haleem, D.J. 2016. Drug targets for obesity and depression: From serotonin to leptin. *Curr Drug Targets* 17(11): 1282–91.

Halford, J.C., E.J. Boyland, C.L. Lawton, J.E. Blundell, and J.A. Harrold. 2011. Serotonergic anti-obesity agents: Past experience and future prospects. *Drugs* 71(17): 2247–55.

Harrold, J.A., and J.C. Halford. 2006. The hypothalamus and obesity. *Recent Pat CNS Drug Discov* 1(3): 305–14.

Heisler, L.K., H.M. Chu, and L.H. Tecott. 1998. Epilepsy and obesity in serotonin 5-HT2C receptor mutant mice. *Ann N Y Acad Sci* 861: 74–8.

Higgins, G.A., and P.J. Fletcher. 2015. Therapeutic potential of 5-HT2C receptor agonists for addictive disorders. *ACS Chem Neurosci* 6(7): 1071–88.

Higgins, G.A., E.M. Sellers, and P.J. Fletcher. 2013. From obesity to substance abuse: Therapeutic opportunities for 5-HT2C receptor agonists. *Trends Pharmacol Sci* 34(10): 560–70.

Higgins, G.A., F.D. Zeeb, and P.J. Fletcher. 2017. Role of impulsivity and reward in the anti-obesity actions of 5-HT2C receptor agonists. *J Psychopharmacol* 31(11): 1403–18.

Hocking, S., A. Dear, and M.A. Cowley. 2017. Current and emerging pharmacotherapies for obesity in Australia. *Obes Res Clin Pract* 11(5): 501–21.

Hurt, R.T., J. Edakkanambeth Varayil, and J.O. Ebbert. 2014. New pharmacological treatments for the management of obesity. *Curr Gastroenterol Rep* 16(6): 394.

Ikarashi, N., T. Toda, Y. Hatakeyama, et al. 2018. Anti-hypertensive effects of acacia polyphenol in spontaneously hypertensive rats. *Int J Mol Sci* 19(3): 700.

Ikarashi, N., T. Toda, T. Okaniwa, K. Ito, W. Ochiai, and K. Sugiyama. 2011. Anti-obesity and anti-diabetic effects of acacia polyphenol in obese diabetic KKAy mice fed high-fat diet. *Evid Based Complement Alternat Med* 2011: 952031.

Izaola, O., D. de Luis, I. Sajoux, J.C. Domingo, and M. Vidal. 2015. Inflamación Y obesidad (lipoinflamación) [Inflammation and obesity (lipoinflammation)]. *Nutr Hosp* 31(6): 2352–8.
Jangra, S., and R. Pothuraju. 2020. Functional significance of gum acacia in the management of obesity. *Curr Pharm Des* 26(3): 293–5.
Jangra, S., R.K. Sharma, R. Pothuraju, and A.K. Mohanty. 2019. Ameliorative effect of fermentable fibres on adiposity and insulin resistance in C57BL/6 mice fed a high-fat and sucrose diet. *Food Funct* 10(6): 3696–705.
Jensen, N.H., T.I. Cremers, and F. Sotty. 2010. Therapeutic potential of 5-HT2C receptor ligands. *ScientificWorldJournal* 10: 1870–85.
Joo, J.K., and K.S. Lee. 2014. Pharmacotherapy for obesity. *J Menopausal Med* 20(3): 90–6.
Karth, M.M., B.J. Baugher, N. Daly, M.D. Karth, S.C. Gironda, and B.D. Sachs. 2019. Brain 5-HT deficiency prevents antidepressant-like effects of high-fat-diet and blocks high-fat-diet-induced GSK3β phosphorylation in the hippocampus. *Front Mol Neurosci* 12: 298.
Kassir, R. 2020. Risk of COVID-19 for patients with obesity. *Obes Rev* 21(6): e13034.
Kesić, M., P. Baković, M. Horvatiček, B.L.J. Proust, J. Štefulj, and L. Čičin-Šain. 2020. Constitutionally high serotonin tone favors obesity: Study on rat sublines with altered serotonin homeostasis. *Front Neurosci* 14: 219.
Khorassani, F.E., A. Misher, and S. Garris. 2015. Past and present of antiobesity agents: Focus on monoamine modulators. *Am J Health Syst Pharm* 72(9): 697–706.
Kim, M., I. Hwang, H.S. Pagire, et al. 2020. Design, synthesis, and biological evaluation of new peripheral 5HT2A antagonists for nonalcoholic fatty liver disease. *J Med Chem* 63(8): 4171–82.
Kim, O.Y., J.Y. Chung, and J. Song. 2019. Effect of resveratrol on adipokines and myokines involved in fat browning: Perspectives in healthy weight against obesity. *Pharmacol Res* 148: 104411.
Kukhetpitakwong, R., C. Hahnvajanawong, P. Homchampa, V. Leelavatcharamas, J. Satra, and W. Khunkitti. 2006. Immunological adjuvant activities of saponin extracts from the pods of *Acacia concinna*. *Int Immunopharmacol* 6(11): 1729–35.
Kurhe, Y., and R. Mahesh. 2015. Mechanisms linking depression co-morbid with obesity: An approach for serotonergic type 3 receptor antagonist as novel therapeutic intervention. *Asian J Psychiatr* 17: 3–9.
Kwaifa, I.K., H. Bahari, Y.K. Yong, and S.M. Noor. 2020. Endothelial dysfunction in obesity-induced inflammation: Molecular mechanisms and clinical implications. *Biomolecules* 10(2): 291.
Lansky, E.S. 2020a. *Review of Joseph Tarfur's Fellowship of the River*. Harefuah.
Lansky, E.S. 2020b. Novel harmino-ocudelic tuning (HOT) for ocular disorders. *Med Hypotheses* 143: 109834.
Lansky, E.S., H.M. Paavilainen, and S. Lansky. 2017. *Harmal: The Genus Peganum*. Boca Raton, FL: CRC Press.
Lee, S.H., G. Paz-Filho, C. Mastronardi, J. Licinio, and M.L. Wong. 2016. Is increased antidepressant exposure a contributory factor to the obesity pandemic? *Transl Psychiatry* 6(3): e759.
Lernmark, A. 1971. The significance of 5-hydroxytryptamine for insulin secretion in the mouse. *Horm Metab Res* 3(5): 305–9.
Li, R.J., Y. Liu, H.Q. Liu, and J. Li. 2020. Ketogenic diets and protective mechanisms in epilepsy, metabolic disorders, cancer, neuronal loss, and muscle and nerve degeneration. *J Food Biochem* 44(3): e13140.
Lin, Y., B. Bao, H. Yin, et al. 2019. Peripheral cathepsin L inhibition induces fat loss in *C. elegans* and mice through promoting central serotonin synthesis. *BMC Biol* 17(1): 93.
Lin, Y., N. Yang, B. Bao, L. Wang, J. Chen, and J. Liu. 2020. Luteolin reduces fat storage in *Caenorhabditis elegans* by promoting the central serotonin pathway. *Food Funct* 11(1): 730–40.
Lyte, J.M., C.E. Gheorghe, M.S. Goodson, et al. 2020. Gut-brain axis serotonergic responses to acute stress exposure are microbiome-dependent. *Neurogastroenterol Motil* 32(11): e13881.
Magalhães, C.P., M.F. de Freitas, M.I. Nogueira, et al. 2010. Modulatory role of serotonin on feeding behavior. *Nutr Neurosci* 13(6): 246–55.
Mangge, H., C. Ciardi, K. Becker, B. Strasser, D. Fuchs, and J.M. Gostner. 2017. Influence of antioxidants on leptin metabolism and its role in the pathogenesis of obesity. *Adv Exp Med Biol* 960: 399–413.
Mangge, H., K. Summers, G. Almer, et al. 2013. Antioxidant food supplements and obesity-related inflammation. *Curr Med Chem* 20(18): 2330–7.
Manning, S., A. Pucci, and N. Finer. 2014. Pharmacotherapy for obesity: Novel agents and paradigms. *Ther Adv Chronic Dis* 5(3): 135–48.
Mao, Z., H. Lin, W. Su, et al. 2019. Deficiency of ZnT8 promotes adiposity and metabolic dysfunction by increasing peripheral serotonin production. *Diabetes* 68(6): 1197–209.
Marks, J.W. (ed.). 2021. Medical definition of psychedelic. *MedicineNet*. www.medicinenet.com/psychedelic/definition.htm.

Marston, O.J., A.S. Garfield, and L.K. Heisler. 2011. Role of central serotonin and melanocortin systems in the control of energy balance. *Eur J Pharmacol* 660(1): 70–9.
Marston, O.J., and L.K. Heisler. 2009. Targeting the serotonin 2C receptor for the treatment of obesity and type 2 diabetes. *Neuropsychopharmacol* 34(1): 252–3.
Martin, A.M., J.M. Yabut, J.M. Choo, et al. 2019. The gut microbiome regulates host glucose homeostasis via peripheral serotonin. *Proc Natl Acad Sci U S A* 116(40): 19802–4.
Martin, A.M., R.L. Young, L. Leong, et al. 2017. The diverse metabolic roles of peripheral serotonin. *Endocrinology* 158(5): 1049–63.
Miller, K.J. 2005. Serotonin 5-HT2C receptor agonists: Potential for the treatment of obesity. *Mol Interv* 5(5): 282–91.
Namkung, J., H. Kim, and S. Park. 2015. Peripheral serotonin: A new player in systemic energy homeostasis. *Mol Cells* 38(12): 1023–8.
Nasir, O. 2013. Renal and extrarenal effects of gum arabic (*Acacia senegal*)—what can be learned from animal experiments? *Kidney Blood Press Res* 37(4–5): 269–79.
Nasir, O., F. Artunc, K. Wang, et al. 2010. Downregulation of mouse intestinal Na(+)-coupled glucose transporter SGLT1 by gum arabic (*Acacia senegal*). *Cell Physiol Biochem* 25(2–3): 203–10.
Nie, T., X. Hui, L. Mao, et al. 2016. Harmine induces adipocyte thermogenesis through RAC1-MEK-ERK-CHD4 Axis. *Sci Rep* 6: 36382.
Nieuwenhuis, D., N. Pujol-Gualdo, I.A.C. Arnoldussen, and A.J. Kiliaan. 2020. Adipokines: A gear shift in puberty. *Obes Rev* 21(6): e13005.
O'Donnell, V.B., K. Ekroos, G. Liebisch, and M. Wakelam. 2020. Lipidomics: Current state of the art in a fast moving field. *Wiley Interdiscip Rev Syst Biol Med* 12(1): e1466.
Ogawa, S., and Y. Yazaki. 2018. Tannins from *Acacia mearnsii* De Wild. bark: Tannin determination and biological activities. *Molecules* 23(4): 837.
Oh, C.M., S. Park, and H. Kim. 2016. Serotonin as a new therapeutic target for diabetes mellitus and obesity. *Diabetes Metab J* 40(2): 89–98.
Palacios, J.M., A. Pazos, and D. Hoyer. 2017. A short history of the 5-HT2C receptor: From the choroid plexus to depression, obesity and addiction treatment. *Psychopharmacol (Berl)* 234(9–10): 1395–418.
Pan, Q., Q. Liu, R. Wan, et al. 2019. Selective inhibition of intestinal 5-HT improves neurobehavioral abnormalities caused by high-fat diet mice. *Metab Brain Dis* 34(3): 747–61.
Patro-Gołąb, B., B.M. Zalewski, M. Kołodziej, et al. 2016. Nutritional interventions or exposures in infants and children aged up to 3 years and their effects on subsequent risk of overweight, obesity and body fat: A systematic review of systematic reviews [correction published in 2018 in *Obes Rev* 19(11): 1620]. *Obes Rev* 17(12): 1245–57.
Peterlin, B.L., A.M. Rapoport, and T. Kurth. 2010. Migraine and obesity: Epidemiology, mechanisms, and implications. *Headache* 50(4): 631–48.
Proietto, J., B.C. Fam, D.A. Ainslie, and A.W. Thorburn. 2000. Novel anti-obesity drugs. *Expert Opin Investig Drugs* 9(6): 1317–26.
Rahm, I., O. Spigset, and T. Mjörndal. 2004. Dieting and weight loss do not affect on the platelet serotonin 5-HT2A receptor. *Eur Neuropsychopharmacol* 14(5): 381–3.
Rebholz, H., E. Friedman, and J. Castello. 2018. Alterations of expression of the serotonin 5-HT4 receptor in brain disorders. *Int J Mol Sci* 19(11): 3581.
Rozenblit-Susan, S., N. Chapnik, and O. Froy. 2019. Serotonin prevents differentiation of brown adipocytes by interfering with their clock. *Obesity (Silver Spring)* 27(12): 2018–24.
Ruangaram, W., and E. Kato. 2020. Selection of Thai medicinal plants with anti-obesogenic potential via in vitro methods. *Pharmaceuticals (Basel)* 13(4): E56.
Ruiz-Fernández, C., V. Francisco, J. Pino, et al. 2019. Molecular relationships among obesity, inflammation and intervertebral disc degeneration: Are adipokines the common link? *Int J Mol Sci* 20(8): 2030.
Saad, M.J., A. Santos, and P.O. Prada. 2016. Linking gut microbiota and inflammation to obesity and insulin resistance. *Physiology (Bethesda)* 31(4): 283–93.
Salisbury, A.L., G.D. Papandonatos, L.R. Stroud, A.K. Smith, and P.A. Brennan. 2020. Prenatal antidepressant exposures and gastrointestinal complaints in childhood: A gut-brain axis connection? *Dev Psychobiol* 62(6): 816–28.
Sargent, B.J., and A.J. Henderson. 2011. Targeting 5-HT receptors for the treatment of obesity. *Curr Opin Pharmacol* 11(1): 52–8.
Serralde-Zúñiga, A.E., A.G. Gonzalez Garay, Y. Rodríguez-Carmona, and G. Melendez. 2019. Fluoxetine for adults who are overweight or obese. *Cochrane Database Syst Rev* 10(10): CD011688.
Sethi, J.K., and A.J. Vidal-Puig. 2007. Targeting fat to prevent diabetes. *Cell Metab* 5(5): 323–5.

Shabbir, F., A. Patel, C. Mattison, et al. 2013. Effect of diet on serotonergic neurotransmission in depression. *Neurochem Int* 62(3): 324–9.
Shettar, V., S. Patel, and S. Kidambi. 2017. Epidemiology of obesity and pharmacologic treatment options. *Nutr Clin Pract* 32(4): 441–62.
Simonyi, G., G. Pados, M. Medvegy, and J.R. Bedros. 2012. Az elhízás gyógyszeres kezelése: múlt, jelen, jövő [The pharmacological treatment of obesity: Past, present and future]. *Orv Hetil* 153(10): 363–73.
Song, J., D. Hanniford, C. Doucette, et al. 2005. Development of homogeneous high-affinity agonist binding assays for 5-HT2 receptor subtypes. *Assay Drug Dev Technol* 3(6): 649–59.
Stolarczyk, E. 2017. Adipose tissue inflammation in obesity: A metabolic or immune response? *Curr Opin Pharmacol* 37: 35–40.
Szőke, H., Z. Kovács, I. Bókkon, et al. 2020. Gut dysbiosis and serotonin: Intestinal 5-HT as a ubiquitous membrane permeability regulator in host tissues, organs, and the brain. *Rev Neurosci* 31(4): 415–25.
Tabrizi, R., O.R. Tamtaji, K.B. Lankarani, et al. 2020. The effects of resveratrol intake on weight loss: A systematic review and meta-analysis of randomized controlled trials. *Crit Rev Food Sci Nutr* 60(3): 375–90.
Tezuka, Y., K. Honda, A.H. Banskota, M.M. Thet, and S. Kadota. 2000. Kinmoonosides A-C, three new cytotoxic saponins from the fruits of *Acacia concinna*, a medicinal plant collected in Myanmar. *J Nat Prod* 63(12): 1658–64.
Vadivel, V., and H.K. Biesalski. 2012. Effect of certain indigenous processing methods on the bioactive compounds of ten different wild type legume grains. *J Food Sci Technol* 49(6): 673–84.
van Galen, K.A., K.W. Ter Horst, J. Booij, S.E. la Fleur, and M.J. Serlie. 2018. The role of central dopamine and serotonin in human obesity: Lessons learned from molecular neuroimaging studies. *Metabolism* 85: 325–39.
Veniaminova, E., R. Cespuglio, I. Chernukha, et al. 2020a. Metabolic, molecular, and behavioral effects of Western diet in serotonin transporter-deficient mice: Rescue by heterozygosity? *Front Neurosci* 14: 24.
Veniaminova, E., M. Oplatchikova, L. Bettendorff, et al. 2020b. Prefrontal cortex inflammation and liver pathologies accompany cognitive and motor deficits following Western diet consumption in non-obese female mice. *Life Sci* 241: 117163.
Vickers, S.P., and C.T. Dourish. 2004. Serotonin receptor ligands and the treatment of obesity. *Curr Opin Investig Drugs* 5(4): 377–88.
Voigt, J.P., and H. Fink. 2015. Serotonin controlling feeding and satiety. *Behav Brain Res* 277: 14–31.
Wang, P., J.C. Alvarez-Perez, D.P. Felsenfeld, et al. 2015. A high-throughput chemical screen reveals that harmine-mediated inhibition of DYRK1A increases human pancreatic beta cell replication. *Nat Med* 21(4): 383–8.
Wu, X., J. Huang, C. Shen, et al. 2020. NRF2 deficiency increases obesity susceptibility in a mouse menopausal model. *PLoS One* 15(2): e0228559.
Wurtman, J., and R. Wurtman. 2018. The trajectory from mood to obesity. *Curr Obes Rep* 7(1): 1–5.
Wurtman, R.J., and J.J. Wurtman. 1995. Brain serotonin, carbohydrate-craving, obesity and depression. *Obes Res* 3(Suppl 4): 477S–80S.
Wyler, S.C., C.C. Lord, S. Lee, J.K. Elmquist, and C. Liu. 2017. Serotonergic control of metabolic homeostasis. *Front Cell Neurosci* 11: 277.
Xu, D., M. Fang, H. Wang, L. Huang, Q. Xu, and Z. Xu. 2020. Enhanced production of 5-hydroxytryptophan through the regulation of L-tryptophan biosynthetic pathway. *Appl Microbiol Biotechnol* 104(6): 2481–8.
Yabut, J.M., J.D. Crane, A.E. Green, D.J. Keating, W.I. Khan, and G.R. Steinberg. 2019. Emerging roles for serotonin in regulating metabolism: New implications for an ancient molecule. *Endocr Rev* 40(4): 1092–107.
Yabut, J.M., E.M. Desjardins, E.J. Chan, et al. 2020. Genetic deletion of mast cell serotonin synthesis prevents the development of obesity and insulin resistance. *Nat Commun* 11(1): 463.
Zhao, Y., B. Chen, J. Shen, et al. 2017. The beneficial effects of quercetin, curcumin, and resveratrol in obesity. *Oxid Med Cell Longev* 2017: 1459497.
Zhuang, P., Y. Zhang, Q. Shou, et al. 2020. Eicosapentaenoic and docosahexaenoic acids differentially alter gut microbiome and reverse high-fat diet-induced insulin resistance. *Mol Nutr Food Res* 64(10): e1900946.

9 Acacias and Diabetes

Diabetes mellitus, or "sweet diabetes," and *diabetes insipidus*, which is a pituitary disorder, are mechanistically unrelated, though both usually involve copious amounts of dilute urine; however, only the first of these urines is sweet. Old-time physicians would judiciously taste a drop of a patient's urine to make the diagnosis of "sugar diabetes." As with inflammation, the etiology is clearer in the Hebrew language, with the word for diabetes mellitus being *sakeret* and the word for sugar being *sukar*: both words with the same sweet root. In modern usage, the word "diabetes" by itself refers to diabetes mellitus.

There are also different types of diabetes mellitus, all of which at some point have the sign of high blood sugar, which may or may not spill over into the urine sufficiently to be detectable in it. Type 1 diabetes, therefore, is a subtype of diabetes mellitus, one that was formerly known as juvenile onset diabetes and which involves the body's own "autoimmune" reaction against its own pancreatic beta cells in the islets of Langerhans. The job of these beta cells includes the manufacture of insulin, and when the beta cells are down, regular subcutaneous or intramuscular injections of insulin are medically considered essential for life. Type 1 diabetes, which can be survived beyond the juvenile period with good care, is the type of diabetes that may be improved by ingestion of the beta carboline indole alkaloid harmine, as previously discussed (Wang et al. 2015). Otherwise, it demands insulin injections and extra care.

The more common type of diabetes mellitus is type 2, formerly known as adult-onset diabetes mellitus, or AODM. Type 2 diabetes accounts for 90% of all diabetes patients. Insulin resistance is the main cause and itself the main cause of metabolic syndrome and obesity. The threads between diabetes, insulin resistance, metabolic syndrome, obesity, chronic inflammation, and schizophrenia, are intertwined and not always easy to tease apart. But diabetes, that is, type 2 diabetes, is characterized by recognizable symptoms: sugar in the urine, excess sugar in the blood, polyuria, polydipsia, and polyphagia, and of course, symptoms of a diabetic emergency may involve changes in consciousness, vertigo, etc. Long-term consequences of diabetes include damage to the blood vessels of retina, brain, heart, and kidney, which lead to different complications and damage to the organism. The symptoms become causes, and *vice versa*, resulting in the interwoven complexity of the disease and its metabolic partners, which threaten to wreck health and destroy life.

Acacias do have something to offer for the treatment of diabetes, and since most of the diabetes in the world is type 2 diabetes, discussion about diabetes and acacias will refer, unless otherwise specified, to type 2 diabetes. Although the antidiabetic virtues and potentialities researched within *Acacia s.l.* extend over a few different species, the majority are from one, *Vachellia nilotica*.

Although *V. nilotica* is sometimes referred to as the "gum arabic tree," it is not to be confused with *Senegalia senegal*, from which the official gum arabic of pharmacy is lawfully derived. *V. nilotica*, rather, is identical to the plant also widely known as *Acacia arabica* or *Vachellia arabica*. In India, where the bark, seed pods, and leaves are used in Ayurvedic medicine, the *V. nilotica* tree is known as *babul*, or "*Israeli babul*" (Kapoor 1990).

The *babul*, *V. nilotica*, was formerly the "type species" for the genus formerly known as *Acacia* (i.e., *Acacia s.l.*). Named *Acacia* by Dioscorides (c. 40–90 CE) for the Greek word meaning "thorned," the specific species name *nilotica* was given by Linnaeus (1707–1778) for the *babul*'s habitat along the River Nile in Egypt. Thus, praised for its medicinal virtues from antiquity, this most valuable and formerly most representative (as type species) of all acacias now finds itself expelled from the former *Acacia* genus, reviled as an invasive weed by both the United States and Australia, while the remaining *Acacia* (*sensu stricto*) species in Australia, the majority of which are thornless,

DOI: 10.1201/9780429440946-9

(a)

(b)

FIGURE 9.1 Bark and leaves of *Vachellia nilotica* contain potentially antidiabetic tannins. (a) *V. nilotica* branch with leaves. (5/3/2021, City of Tshwane Metropolitan Municipality, South Africa: by Hildegard Klein, CC BY 4.0, via iNaturalist, www.inaturalist.org/photos/125963762.) (b) *V. nilotica* tree trunk, bark. (Photo by Ashish D. Panchal, via Shutterstock, www.shutterstock.com/image-photo/beautiful-texture-wooden-background-close-brown-1803601909.)

are now viewed as the "true acacias," and only they have retained the genus name of *Acacia*. Nevertheless, the medicinal virtues of *V. nilotica* have been the subject of considerable modern research, of which no small portion is related to the *babul*'s value for diabetes.

One of the "Israeli *babul*" studies of *Acacia* is notable for its selection of compounds aimed to "standardize" or control the composition of successive batches of *V. nilotica* bark extracts. To wit, the compounds cited by the illustrious Mosab Yahya Al-Nour, PhD, that work synergistically within a complex extract of Israeli *babul* to produce ideal therapeutic effects were kaempferol, ellagic acid, and quercetin (Al-Nour et al. 2019). These were known from previous work with pomegranate fruit, where kaempferol and quercetin (flavonols and a flavone) were associated with strong and measurable estrogenic activity, and secondly, ellagic acid plus luteolin plus punicic acid within *Punica granatum* produced therapeutic synergy against inflammation (Lansky et al. 2005a, b). In all respects, Al-Nour's direction for *V. nilotica* medicinally dovetailed with earlier work with *P. granatum* for establishing an appropriate model for complexity, with overlap between *P. granatum* and *V. nilotica*, common tannin, and flavonoid control points within a single plant, an herbal "simple" (Lansky and Von Hoff 2005).

Streptozotocin is a naturally occurring antibiotic molecule originally discovered in the 1950s by Upjohn (Kalamazoo, MI, USA) scientists in a soil bacterium, *Streptomyces achromogenes*. Because of its high toxicity to pancreatic beta cells, streptozotocin is used both as a specialized form of end stage cancer therapy and for creating animal models for diabetes research. Some of the studies to follow used streptozotocin-induced diabetic animals to assess the potential antidiabetic action of acacias.

For example, in a study of 120 rats, 90 were given streptozotocin intraperitoneally in order to induce experimental diabetes. A fraction of these animals also received a single dose of a *V. nilotica* leaf extract. Blood glucose, serum insulin, triglycerides, and platelet aggregation were significantly lower in the rats treated with the Israeli *babul* leaf extract than in the negative control group (Asad, Aslam et al. 2011; Asad, Munir et al. 2011).

In another similar study, diabetes was induced in rats by the injection of another compound, *alloxan*. A group of 12 diabetic rats and 12 non-diabetic control rats, received a regular oral dose of an extract of *V. nilotica* fruits over 5 weeks. The control rats that received the acacia extract showed significantly reduced blood sugar after 3 weeks ($p < 0.005$), but the serum glucose levels of the diabetic rats were not significantly affected. However, the diabetic rats that received the extract did show significantly lower levels of triglycerides ($p < 0.02$) and low-density lipoprotein cholesterol ($p < 0.001$), illustrating the likely benefits of the extract for patients with diabetes (Abuelgassim 2013). An additional study in Islamabad revealed significant dose-dependent decreases in fasting blood glucose, total cholesterol, triglyceride, phospholipids, low-density lipoprotein, and very-low-density lipoprotein and an increase in high-density lipoprotein and serum insulin in alloxan-induced diabetic rats treated with an *V. nilotica* leaf extract, which were comparable to those observed in rats that were treated with the conventional hypoglycemic drug, glibenclamide (Asad et al. 2015).

Traditional African use of *V. nilotica* for diabetes inspired a Nigerian investigation of *babul* chemistry. Aqueous, ethyl acetate, and N-butanol fractions of *V. nilotica* pod (fruit) revealed tannins, saponins, flavonoids, and carbohydrate. The residue remains after all the extractions contained only carbohydrate and tannins. Auwal et al. (2014) used flame emission and atomic absorption spectrophotometry to identify cadmium, calcium, copper, iron, magnesium, manganese, phosphorous, potassium, sodium, and zinc in the pod residue.

In another study in 36 all-female albino rats at the King Fahd Research Center, King Abdulaziz University in Jeddah, Saudi Arabia, diabetes was induced with streptozotocin in half the animals. Of those, three groups of six animals each received either no treatment or 100 mg/kg or 200 mg/kg extract of *V. nilotica*. Dose-dependently, animals receiving the extract exhibited significant decreases in serum glucose, calculated insulin resistance, total cholesterol, triglycerides, low-density lipoprotein C, and *malondialdehyde*, with significant increases in high-density lipoprotein C and coenzyme Q10 compared with the untreated diabetic animals (Hegazy et al. 2013).

In another set of experiments at the National Research Center in Cairo, Egypt, employing an aqueous methanolic extract of *V. nilotica* pods, investigators focused on renal pathology in streptozotocin-induced diabetic rats. The diabetic animals displayed elevated serum urea and creatine as well as hyperglycemia. Diabetic kidney pathology revealed increased lipid peroxidation and superoxide dismutase, as well as decreased glutathione. Histopathological examination exhibited infiltration of the lymphocytes into the interstitial spaces, glomerular hypertrophy, thickening of the basement membrane, tubular necrosis, and brush border loss in a portion of the proximal convoluted tubules. Omara et al. (2012) found that treatment with *V. nilotica* extracts stopped or reversed most of this pathology and partially fixed the structure of the kidneys.

Detailed guidelines for an in-depth clinical evaluation of *V. nilotica* extracts as therapy for type 2 diabetes were created by a cooperation between researchers at two different Iranian universities, Hormozgan University of Medical Sciences, Bandar Abbas, Iran, and Shahid Beheshti University of Medical Sciences, Tehran. The authors suggested evaluating subjects in the studies by fasting blood sugar, glucose challenge test, glucose tolerance test, and an oral glucose test obtained by checking 2-hour postprandial glucose and glycated hemoglobin (HbA1c) levels. Further, the authors noted compounds known to occur in *V. nilotica*, i.e., *catechin, galloylated favan-3,4-diol, robidandiol, androstene steroid, d-pinitol carbohydrate, catechin-5-galloyl ester*, and *gallic, m-digallic*, and *chlorogenic* acids (Roozbeh et al. 2017).

A study from the University of North Bengal and North Bengal Medical College in India set out to justify the antidiabetic use of *V. nilotica* in Bangladesh, India, Pakistan, Egypt, and Nigeria as well as its traditional use in Ayurveda. Working with alloxanized mice (blood sugar >/= 200 mg/kg), a *V. nilotica* extract was administered orally at 50 or 200 mg/kg and classical parameters measured after 20 days. The *V. nilotica* treatment resulted in a 65% decrease ($p < 0.001$) in systemic glucose load in the diabetic mice from 398% pretreatment to 65% post-treatment, also lowering insulin resistance by 35% without affecting insulin sensitivity ($p > 0.05$) and decreasing glycosylated hemoglobin HbA1c (34%; $p < 0.001$). Overall, an antioxidant effect was observed, with mitigation of renal and hepatic injury and normalization of lipidemia that followed administration of *V. nilotica* extract to diabetic mice. Strong binding by flavonoids, tocopherols, and sterols within the extract to Nrf2 protein suggested "crosstalk" between the extract and intracellular antioxidant defense mechanisms (Saha et al. 2018).

Inclusion of seed meal of *V. nilotica* or of *Senegalia catechu*, *S. polyacantha*, *S. modesta*, *Acacia benthamii*, or *A. melanoxylon* in the diets of young normal albino mice, resulted in a significant reduction in their glycemias (Singh et al. 1975, 1976, 1977). Powdered seeds of *V. nilotica* administered in doses of 2–4 g/kg to normal rabbits caused a dose-dependent hypoglycemic effect but had no hypoglycemic effect in alloxan-induced diabetic rabbits (Wadood et al. 1989). In the db/db genetically modified mouse model of type 2 diabetes mellitus, of the 10 natural products considered most promising from an *in vitro* screen of 203 commercially available products representing 90 botanical species, *V. nilotica* bark extract was unique in inhibiting both non-fasting glucose and non-fasting insulin (Babish et al. 2010). In a screen of 18 plants with ethnographic use in Sudan consistent with possible anti diabetes action, only *V. nilotica* and *Zizyphus spina-christi* yielded extracts that potently inhibited both alpha amylase (an enzyme that breaks complex carbohydrates into glucose) and lipase (an enzyme that catalyzes the hydrolysis of fats), further substantiating the breadth of the Israeli *babul*'s antidiabetic potency (Elbashir et al. 2018). In a study employing a chloroform extract of *V. nilotica* bark, administered at 250 or 500 mg/kg to alloxan-induced albino rats, decreased blood sugar was coincident with reduced total cholesterol, triglycerides, and LDL, with an increase in HDL, all to pre-diabetic values (Patil et al. 2011).

In a screening of Nigerian plants by Nigerian scientists, *V. nilotica* crude methanolic extract was shown to inhibit enzyme protein tyrosine phosphatase 1B (PTP 1B), an identified drug target for type 2 diabetes, by 55%, which was better than that reported for the standard inhibitor, sumarin (Saidu et al. 2016). A combined extract of five parts *V. nilotica* and 1 part *Humulus lupulus* to db/db diabetic mice (100 mg/kg) for 7 days resulted in a 22% decrease in non-fasting glucose and a 19% decrease in insulin, comparable to that of 0.5 mg/kg rosiglitazone and better than 100 mg/kg metformin (Tripp et al. 2012).

In that sprouting and oil-frying of *V. nilotica* seeds resulted in a significant increase in total free phenolics (9–27%) and tannins (12–28%) but had a diminishing effect on phytic acid and l-DOPA, it was considered an ideal and renewable means of utilizing the Israeli *babul* as a medicinal food (Vadivel and Biesalski 2012).

Thirty-six different endophytic fungi were isolated from *V. nilotica* to screen for inhibitors of amylase or α-glucosidase. One isolate of *Aspergillus awamori* significantly inhibited both (Singh and Kaur 2016).

"The only thing we knew for sure about Henry Porter was that his name wasn't Henry Porter" (Dylan 1986).

The tree (*V. nilotica*) known as the "gum arabic tree" is not the tree whose gummy stem exudate is legally "gum arabic." Real gum arabic is from *Senegalia senegal*, though an exudate of *Vachellia seyal* is often included. The exudate of the gum arabic tree is not gum arabic. The *Acacia s.l.* "type species," *V. nilotica* (*Acacia arabica*), is not *Acacia sensu stricto*.

Gum arabic is a hugely important article in the pharmacist's tool kit for making emulsions of many kinds, facilitating the preparation of medicaments for oral and also parenteral applications.

(a) (b) (c) (d)

FIGURE 9.2 *Acacia melanoxylon*, "Australian blackwood," is so named for its dark chocolate brown heartwood, which, when aged, is almost black owing to its rich tannin concentration. The wood is prized for furniture and other fine woodwork. Indigenous Australians made an analgesic compound from the wood as well. The tree is considered an invasive pest in South Africa, California, and the Portuguese Azore Islands. (a) *A. melanoxylon* branch with flowers and leaves. (8/20/2019, Moruya Heads NSW 2537, Australia: by Jenny, CC0 1.0, via iNaturalist, www.inaturalist.org/photos/48728680.) (b) *A. melanoxylon*, close-up of flowers. (4/8/2009, MISC HQ Piiholo, Maui, Hawaii: by Forest & Kim Starr, CC BY 4.0, via Starr Environmental, www.starrenvironmental.com/images/image/?q=24857893221.) (c) *A. melanoxylon*, mature seed pods and seeds. (5/30/2009, Olinda VIC Australia: by John Tann, CC BY 2.0, via Flickr, www.flickr.com/photos/31031835@N08/3589033502/.) (d) *A. melanoxylon* tree growing on the right side of the path. (12/21/2004, Polipoli Trail, Polipoli, Maui, Hawaii: by Forest & Kim Starr, CC BY 4.0, via Starr Environmental, www.starrenvironmental.com/images/image/?q=24354184519.)

Gum arabic is a complex mixture of glycoproteins and polysaccharides. The name "gum arabic" is a translation from the Arabic *al-samgh al-'arabi* referring to the very same gum, a trade article from Sudan since at least 900 CE, with Sudan still supplying 80% of the world's gum arabic (Van Dalen 2020). Gum arabic is water-soluble, and it is used in the soft drink industry as a stabilizer.

FIGURE 9.3 *Cygnus cygnus*, "whooper swan," is the type species of the genus *Cygnus*. (7/17/2022, River Taff Mouth/Cardiff Bay, Cardiff, UK: by Mike Brady, CC BY 4.0, via iNaturalist, www.inaturalist.org/photos/215012735.)

In one investigation of alloxan-induced type 1 diabetic rats, the addition of 15% gum arabic to the rats' drinking water significantly mitigated diabetes-related pathology. The gum arabic-induced therapeutic changes in the diabetic rats included:

- ↑ Activity of enzymes superoxide dismutase, catalase, and glutathione peroxidase resulting in;
- ↑ *Glutathione*
- ↑ Heat shock protein 70, glutathione peroxidase, and superoxide dismutase mRNA
- ↓ Malondialdehyde
- ↓ Alanine transaminase
- ↓ Aspartate transaminase

Ahmed et al. (2015) found that gum arabic may protect the livers of diabetic patients by regulating genes that respond to oxidative stress and improving the body's antioxidant status.

Renal insufficiency of various types and degrees is one of the possible complications of diabetes. In another classic study, diabetes was first induced in rats with a single injection of streptozotocin. Once the diabetes had very rapidly taken hold, some of the rats also began to receive in their drinking water the nucleic acid base, pure adenine, 0.25% w/w for 5 weeks, which, administered in this way, readily creates chronic renal disease in the newly diabetic rats. Some of the rats received adenine alongside the gum arabic in their drinking water at a 15% rate. In the gum arabic–treated animals, the renal disease was significantly mitigated (Al Za'abi et al. 2018).

The visceral adiposity index and hypertriglyceridemic waist phenotype (the simultaneous presence of waist circumference (WC) ≥ 90/80 cm and plasma triglyceride (TG) concentration ≥ 1.7 mmol/L (–1) in both men and women) (Du et al. 2014) provided a useful metric for a clinical trial administering gum arabic to patients with diabetes. Ninety-one patients with diabetes, all on hypoglycemic drugs, age average 50, 73 females and 18 males, participated, daily consuming either 30 g gum arabic or 5 g placebo for 3 months. Those taking the gum arabic compared to the placebo group showed a 2% decrease in body mass index and a 24% decrease in the visceral adiposity index, recognized as a superior predictor of cardiovascular complications of diabetes compared to other indices. Systolic blood pressure also significantly dropped in the group treated with gum arabic, compared to the placebo subjects (Babiker et al. 2018).

Akita mice [akita(+/–)], which spontaneously develop insulin deficiency and thus hyperglycemia, were employed in a study to examine the effects of gum arabic, "a $Ca^{(2+)}$-, $Mg^{(2+)}$- and $K^{(+)}$-rich dietary fiber used in the Middle East for centuries for treating chronic kidney disease," on renal function. A regimen of 10% gum arabic in drinking water "significantly increased urinary excretion of $Ca^{(2+)}$ and significantly decreased plasma phosphate and urea concentrations, urinary flow rate, urinary $Na^{(+)}$, phosphate and glucose excretion, blood pressure, and proteinuria" in the diabetic rodents (Nasir et al. 2012). Further work in the diabetic mouse revealed gum arabic to decrease expression of intestinal Na(+) coupled *glucose carrier SGLT1*, with subsequent delay of electrogenic intestinal glucose transport, glucose-induced hyperglycemia, hyperinsulinemia, and body weight gain, as well as protection against diabetes-associated colon cancer. The findings solidified the rationale for the use of gum arabic in the complementary treatment of diabetes (Nasir 2013).

Earlier, it was mentioned that sprouting and oil-frying acacia seeds was a means of preserving and utilizing their antidiabetic principles; the example given was the seeds of *V. nilotica*. Acacia seed flour is another method that may also provide good results in the treatment of patients, with acacia flour as a medicinal food. In addition to sprouting and frying, the seeds may be allowed to dry and then be ground into flour. The flour can then be mixed with wheat or other grain flour to make a functional anti diabetes bread, as was the case in a study in which the acacia flour was derived from the seeds of *Acacia coriacea*, and mixed with wheat flour according to a ratio of 18 parts *A. coriacea* flour to 82 parts wheat flour. In the study, six healthy human volunteers ate such bread, and significant reductions in their immediate postprandial plasma glucose and insulin at 60 and 90 minutes were noted (Thorburn et al. 1987).

Another Australian acacia with exceptional antidiabetic activity is *Acacia auriculiformis*. Extracts of its stem bark and empty seed pods were tested for anti diabetes activity in alloxan-induced rodents at 200 or 400 mg/kg for bark extracts and 100 or 200 mg/kg for empty seed pod extracts. The treatments with *A. auriculiformis* extracts restored serum glucose, lipids, and renal function indices to pre-alloxan values in parity with the standard antidiabetic drug glibenclamide (Sathya and Siddhuraju 2012, 2013). Extraction of *A. auriculiformis* with non-polar solvents worked better than with polar solvents, and acidified solvents were preferable in creating an extract that was active as an acetylcholinesterase drug, relevant to the treatment and prevention of Alzheimer's disease, which is often comorbid with diabetes (Amir Rawa et al. 2019).

Australian *Acacia ligulata*, in a screen of 12 plant extracts, inhibited a key enzyme in carbohydrate metabolism, *α-glucosidase*, which catalyzes the breakdown of starch and disaccharides to glucose, at an IC(50) of 1.01 μg/mL (Gulati et al. 2012). In another study by the same team, of 7 Australian indigenous people's plant extracts tested, only those of *Acacia kempeana* and *Santalum spicatum* stimulated glucose uptake in adipocytes. Further, the extract of *Acacia tetragonophylla* reduced lipid accumulation in differentiated adipocytes (Gulati et al. 2015).

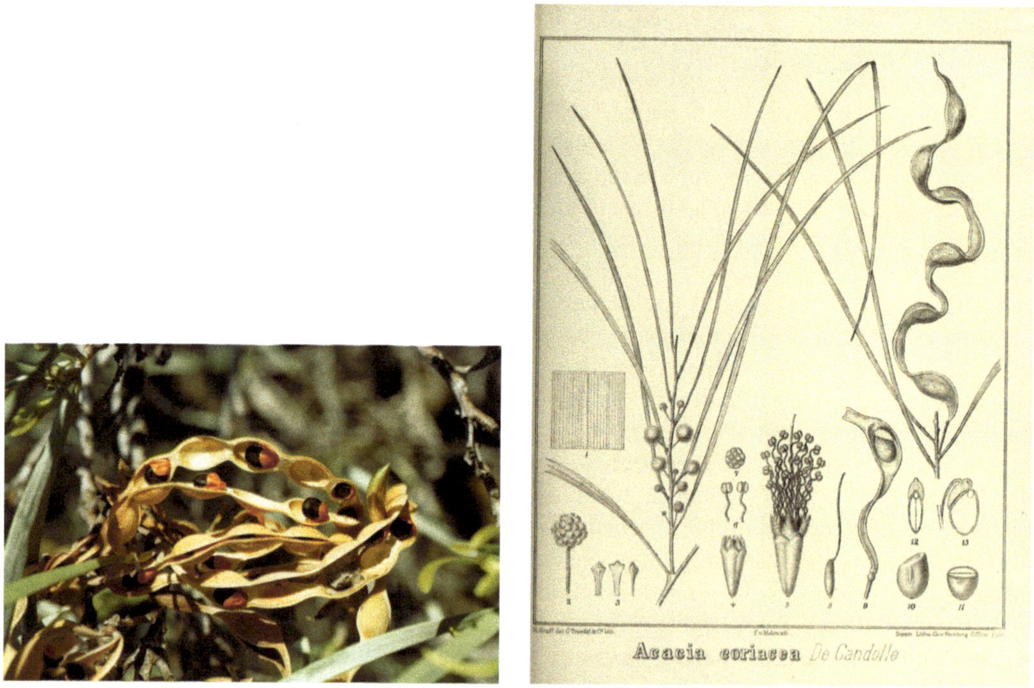

(a) (b)

FIGURE 9.4 *Acacia coriacea*, "river jam," "wirewood," "wiry wattle," or by Indigenous Australians, *gunandru*, *A. coriacea* also sometimes goes by the common and easily confused with other species name "dogwood." (a) *A. coriacea* seeds in open seed pods. (Photo by Maurice MacDonald, CC BY 3.0, via ScienceImage, CSIRO, www.scienceimage.csiro.au/image/1564.) (b) *A. coriacea*, leaves, flowers, seed pods, and seeds; botanical drawing. (In F. J. H. von Mueller, *Iconography of Australian Species of* Acacia *and Cognate Genera*, Table 56 [Melbourne: J. Ferres, Govt. Printer, 1887–88], via www.plantillustrations.org, public domain, www.plantillustrations.org/illustration.php?id_illustration=210440.)

(a) (b)

FIGURE 9.5 *Acacia tetragonophylla*, *curara* or *kurara*, also known as "dead finish," is a largish shrub around 1.5–5 m tall, endemic to semi-arid central and Western Australia. The hard, sharp mature phyllodes may look and prick like thorns, but they are not. With phyllodes like that, you don't need thorns! The seeds may be ground into seed cakes eaten by the indigenous peoples, and the sharp phyllodes inserted into warts to cure them. (a) *A. tetragonophylla* phyllodes. (11/10/2014, Living Desert State Park, Broken Hill, NSW, Australia: by John Tann, CC BY 2.0, via Flickr, www.flickr.com/photos/31031835@N08/15301608233/.) (b) *A. tetragonophylla*, freshly opened seed pod. (11/10/2014, Living Desert State Park, Broken Hill, NSW, Australia: by John Tann, CC BY 2.0, via Flickr, www.flickr.com/photos/31031835@N08/15735508607/.)

Working with another diabetic mouse model, the KKAy, which shows insulin resistance, dislipidemia, and obesity as well as diabetes, the addition of an extract of seed pods from *Acacia mearnsii* to the high-fat murine diet for 7 weeks totally prevented the untoward abnormalities of body weight, plasma glucose, and insulin observed in the high-fat diet controls (Ikarashi et al. 2011). A subsequent study by the same team studied systolic and diastolic blood pressure in both spontaneously hypertensive and normal control Wistar Kyoto rats. Administration of the polyphenol component of *A. mearnsii*, a tree giving sweet shoots for the consumption of Australian indigenous peoples and wood for European wine curing, to rats resulted in significantly higher levels of superoxide dismutase and lower systolic and diastolic blood pressure in the spontaneously hypertensive rats than in the normal ones (Ikarashi et al. 2018).

Both the condensed but especially the hydrolysable tannin of *A. mearnsii* effectively suppressed pancreatic α-amylase with IC(50) 141.1 micromolar (μM) for the hydrolysable and 248.1 μM for the condensed tannin (Kato et al. 2017). In 34 non-diabetic subjects with impaired glucose tolerance, daily intake of *A. mearnsii* tannins over 8 weeks resulted in a robust statistical improvement over placebo in reducing glucose and insulin levels, which gradually declined in the *A. mearnsii* group but not in the placebo controls over the 8-week period, illustrating the potential for this preparation for long-term preventive or ongoing human antidiabetic therapeutics (Ogawa et al. 2013).

Another acacia with strong antidiabetic potential based on ethnographic usage is *Senegalia catechu* (Banjari et al. 2019), which was also hepatoprotective *in vivo* (Hiraganahalli et al. 2012). An aqueous extract of *S. catechu* heartwood significantly reduced blood sugar in glucose-induced hyperglycemic mice with an optimum dose of 400 mg/kg (Rahmatullah et al. 2013). *S. catechu* had one of the highest fidelity ratings as a diabetes treatment in an Indian survey of ethnomedical usage of 197 plants and 112 informants, 78 males and 34 females (Rao et al. 2015).

Strong inhibition of α-glucosidase activity was discovered from a polysaccharide isolated from the gum of *Vachellia tortilis* (Bisht et al. 2013).

A *Vachellia leucophloea* extract inhibited α-amylase =/> 28% (Gautam et al. 2012).

Leaves of *Vachellia seyal* are another source of antidiabetic potential, according to ethnographic usage (Hilmi et al. 2014).

The ethanol-soluble fraction of an aqueous extract of the aerial parts of *Vachellia farnesiana* increased glucose uptake from 40 μg/mL in isolated rat hemidiaphragm *ex vivo* (Kingsley et al. 2014).

Working *in vitro* with *Mus* normal C2C1 myocytes and 3T3-L1 adipocytes, a defatted acetone extract of "sweet thorn," *Vachellia karroo*, was found by researchers at the University of Limpopo, South Africa, to increase translocation of the insulin regulated glucose transporter protein, GLUT4, at 25 μg/mL, with α-amylase inhibition (IC(50) 30.2 ± 3.037 μg/mL) (Njanje et al. 2017).

REFERENCES

Abuelgassim, A.O. 2013. Effect of *Acacia nilotica* fruit extract on serum glucose and lipid concentrations in alloxan-induced diabetic rats. *Pak J Biol Sci* 16(21): 1398–402.

Ahmed, A.A., J.S. Fedail, H.H. Musa, A.A. Kamboh, A.Z. Sifaldin, and T.H. Musa. 2015. Gum Arabic extracts protect against hepatic oxidative stress in alloxan induced diabetes in rats. *Pathophysiology* 22(4): 189–94.

Al-Nour, M.Y., M.M. Ibrahim, and T. Elsaman. 2019. Ellagic acid, kaempferol, and quercetin from *Acacia nilotica*: Promising combined drug with multiple mechanisms of action. *Curr Pharmacol Rep* 5(4): 255–80.

Al Za'abi, M., S. Al Salam, Y. Al Suleimani, P. Manoj, A. Nemmar, and B.H. Ali. 2018. Gum acacia improves renal function and ameliorates systemic inflammation, oxidative and nitrosative stress in streptozotocin-induced diabetes in rats with adenine-induced chronic kidney disease. *Cell Physiol Biochem* 45(6): 2293–304.

Amir Rawa, M.S., Z. Hassan, V. Murugaiyah, T. Nogawa, and H.A. Wahab. 2019. Anti-cholinesterase potential of diverse botanical families from Malaysia: Evaluation of crude extracts and fractions from liquid-liquid extraction and acid-base fractionation. *J Ethnopharmacol* 245: 112160.

Asad, M., M. Aslam, T.A. Munir, A. Nadeem. 2011. Effect of *Acacia nilotica* leaves extract on hyperglycaemia, lipid profile and platelet aggregation in streptozotocin induced diabetic rats. *J Ayub Med Coll Abbottabad* 23(2): 3–7.

Asad, M., T.A. Munir, and N. Afzal. 2011. *Acacia nilotica* leave extract and glyburide: Comparison of fasting blood glucose, serum insulin, beta-thromboglubulin levels and platelet aggregation in streptozotocin induced diabetic rats. *J Pak Med Assoc* 61(3): 247–51.

Asad, M., T.A. Munir, S. Farid, M. Aslam, and S.S. Shah. 2015. Duration effect of *Acacia nilotica* leaves extract and glibenclamide as hypolipidaemic and hypoglycaemic activity in alloxan induced diabetic rats. *J Pak Med Assoc* 65(12): 1266–70.

Auwal, M.S., S. Saka, I.A. Mairiga, K.A. Sanda, A. Shuaibu, and A. Ibrahim. 2014. Preliminary phytochemical and elemental analysis of aqueous and fractionated pod extracts of *Acacia nilotica* (Thorn mimosa). *Vet Res Forum* 5(2): 95–100.

Babiker, R., K. Elmusharaf, M.B. Keogh, and A.M. Saeed. 2018. Effect of gum Arabic (*Acacia Senegal*) supplementation on visceral adiposity index (VAI) and blood pressure in patients with type 2 diabetes mellitus as indicators of cardiovascular disease (CVD): A randomized and placebo-controlled clinical trial. *Lipids Health Dis* 17(1): 56.

Babish, J.G., L.M. Pacioretty, J.S. Bland, D.M. Minich, J. Hu, and M.L. Tripp. 2010. Antidiabetic screening of commercial botanical products in 3T3-L1 adipocytes and db/db mice. *J Med Food* 13(3): 535–47.

Banjari, I., A. Misir, M. Pavlić, P.N. Herath, and V.Y. Waisundara. 2019. Traditional herbal medicines for diabetes used in Europe and Asia: Remedies from Croatia and Sri Lanka. *Altern Ther Health Med* 25(3): 40–52.

Bisht, S., R. Kant, and V. Kumar. 2013. α-D-Glucosidase inhibitory activity of polysaccharide isolated from *Acacia tortilis* gum exudate. *Int J Biol Macromol* 59: 214–20.

Du, T., X. Sun, R. Huo, and X. Yu. 2014. Visceral adiposity index, hypertriglyceridemic waist and risk of diabetes: The China health and nutrition survey 2009. *Int J Obes (Lond)* 38(6): 840–7.

Dylan, B. 1986. Brownsville girl. In: *Knocked Out Loaded*. New York: Columbia Records.

Elbashir, S.M.I., H.P. Devkota, M. Wada, et al. 2018. Free radical scavenging, α-glucosidase inhibitory and lipase inhibitory activities of eighteen Sudanese medicinal plants. *BMC Complement Altern Med* 18(1): 282.

Gautam, B., V. Vadivel, W. Stuetz, and H.K. Biesalski. 2012. Bioactive compounds extracted from Indian wild legume seeds: Antioxidant and type II diabetes-related enzyme inhibition properties. *Int J Food Sci Nutr* 63(2): 242–5.

Gulati, V., P. Gulati, I.H. Harding, and E.A. Palombo. 2015. Exploring the anti-diabetic potential of Australian Aboriginal and Indian Ayurvedic plant extracts using cell-based assays. *BMC Complement Altern Med* 15: 8.

Gulati, V., I.H. Harding, and E.A. Palombo. 2012. Enzyme inhibitory and antioxidant activities of traditional medicinal plants: Potential application in the management of hyperglycemia. *BMC Complement Altern Med* 12: 77.

Hegazy, G.A., A.M. Alnoury, and H.G. Gad. 2013. The role of *Acacia Arabica* extract as an antidiabetic, antihyperlipidemic, and antioxidant in streptozotocin-induced diabetic rats. *Saudi Med J* 34(7): 727–33.

Hilmi, Y., M.F. Abushama, H. Abdalgadir, A. Khalid, and H. Khalid. 2014. A study of antioxidant activity, enzymatic inhibition and in vitro toxicity of selected traditional Sudanese plants with anti-diabetic potential. *BMC Complement Altern Med* 14: 149.

Hiraganahalli, B.D., V.C. Chinampudur, S. Dethe, et al. 2012. Hepatoprotective and antioxidant activity of standardized herbal extracts. *Pharmacogn Mag* 8(30): 116–23.

Ikarashi, N., T. Toda, Y. Hatakeyama, et al. 2018. Anti-hypertensive effects of acacia polyphenol in spontaneously hypertensive rats. *Int J Mol Sci* 19(3): 700.

Ikarashi, N., T. Toda, T. Okaniwa, K. Ito, W. Ochiai, and K. Sugiyama. 2011. Anti-obesity and anti-diabetic effects of acacia polyphenol in obese diabetic KKAy mice fed high-fat diet. *Evid Based Complement Alternat Med* 2011: 952031.

Kapoor, L.D. 1990. *CRC Handbook of Ayurvedic Medicinal Plants*. Boca Raton, FL: CRC Press.

Kato, C.G., G.A. Gonçalves, R.A. Peralta, et al. 2017. Inhibition of α-amylases by condensed and hydrolysable tannins: Focus on kinetics and hypoglycemic actions. *Enzyme Res* 2017: 5724902.

Kingsley, B., S. Kayarohanam, P. Brindha, and A. Subramoniam. 2014. Pharmacodynamic studies on the isolated active fraction of *Acacia farnesiana* (L.) Willd. *Pharmacogn Mag* 10(Suppl 2): S283–S287.

Lansky, E.P., G. Harrison, P. Froom, and W.G. Jiang. 2005a. Pomegranate (*Punica granatum*) pure chemicals show possible synergistic inhibition of human PC-3 prostate cancer cell invasion across Matrigel [correction published in 2005 in *Invest New Drugs* 23(4): 379]. *Invest New Drugs* 23(2): 121–2.

Lansky, E.P., W. Jiang, H. Mo, et al. 2005b. Possible synergistic prostate cancer suppression by anatomically discrete pomegranate fractions. *Invest New Drugs* 23(1): 11–20.

Lansky, E.P., and D.D. Von Hoff. 2005. Complex and simple. *Leuk Res* 29(6): 601–2.
Njanje, I., V.P. Bagla, B.K. Beseni, et al. 2017. Defatting of acetone leaf extract of *Acacia karroo* (Hayne) enhances its hypoglycaemic potential. *BMC Complement Altern Med* 17(1): 482.
Nasir, O. 2013. Renal and extrarenal effects of gum arabic (*Acacia senegal*)—What can be learned from animal experiments? *Kidney Blood Press Res* 37(4–5): 269–79.
Nasir, O., A.T. Umbach, R. Rexhepaj, et al. 2012. Effects of gum arabic (*Acacia senegal*) on renal function in diabetic mice. *Kidney Blood Press Res* 35(5): 365–72.
Ogawa, S., T. Matsumae, T. Kataoka, Y. Yazaki, and H. Yamaguchi. 2013. Effect of acacia polyphenol on glucose homeostasis in subjects with impaired glucose tolerance: A randomized multicenter feeding trial. *Exp Ther Med* 5(6): 1566–72.
Omara, E.A., S.A. Nada, A.R. Farrag, W.M. Sharaf, and S.A. El-Toumy. 2012. Therapeutic effect of *Acacia nilotica* pods extract on streptozotocin-induced diabetic nephropathy in rat. *Phytomedicine* 19(12): 1059–67.
Patil, R.N., R.Y. Patil, B. Ahirwar, and D. Ahirwar. 2011. Evaluation of antidiabetic and related actions of some Indian medicinal plants in diabetic rats. *Asian Pac J Trop Med* 4(1): 20–3.
Rahmatullah, M., M. Hossain, A. Mahmud, et al. 2013. Antihyperglycemic and antinociceptive activity evaluation of 'khoyer' prepared from boiling the wood of *Acacia catechu* in water. *Afr J Tradit Complement Altern Med* 10(4): 1–5.
Rao, P.K., S.S. Hasan, B.L. Bhellum, and R.K. Manhas. 2015. Ethnomedicinal plants of Kathua district, J&K, India. *J Ethnopharmacol* 171: 12–27.
Roozbeh, N., L. Darvish, and F. Abdi. 2017. Hypoglycemic effects of *Acacia nilotica* in type II diabetes: A research proposal. *BMC Res Notes* 10(1): 331.
Saha, M.R., P. Dey, I. Sarkar, et al. 2018. *Acacia nilotica* leaf improves insulin resistance and hyperglycemia associated acute hepatic injury and nephrotoxicity by improving systemic antioxidant status in diabetic mice. *J Ethnopharmacol* 210: 275–86.
Saidu, Y., S.A. Muhammad, A.Y. Abbas, A. Onu, I.M. Tsado, and L. Muhammad. 2016. In vitro screening for protein tyrosine phosphatase 1B and dipeptidyl peptidase IV inhibitors from selected Nigerian medicinal plants. *J Intercult Ethnopharmacol* 6(2): 154–7.
Sathya, A., and P. Siddhuraju. 2012. Role of phenolics as antioxidants, biomolecule protectors and as antidiabetic factors—Evaluation on bark and empty pods of *Acacia auriculiformis*. *Asian Pac J Trop Med* 5(10): 757–65.
Sathya, A., and P. Siddhuraju. 2013. Protective effect of bark and empty pod extracts from *Acacia auriculiformis* against paracetamol intoxicated liver injury and alloxan induced type II diabetes. *Food Chem Toxicol* 56: 162–70.
Singh, B., and A. Kaur. 2016. Antidiabetic potential of a peptide isolated from an endophytic *Aspergillus awamori*. *J Appl Microbiol* 120(2): 301–11.
Singh, K.N., and V. Chandra. 1977. Hypoglycaemic and hypocholesterolaemic effects of proteins of *Acacia milanoxylon* and *Bauhinia retusa* wild leguminous seeds in young albino rats. *J Indian Med Assoc* 68(10): 201–3.
Singh, K.N., V. Chandra, and K.C. Barthwal. 1975. Letter to the editor: Hypoglycaemic activity of *Acacia arabica*, *Acacia benthami* and *Acacia modesta* leguminous seed diets in normal young albino rats. *Indian J Physiol Pharmacol* 19(3): 167–8.
Singh, K.N., R.K. Mittal, and K.C. Barthwal. 1976. Hypoglycaemic activity of *Acacia catechu*, *Acacia suma*, and *Albizzia odoratissima* seed diets in normal albino rats. *Indian J Med Res* 64(5): 754–7.
Thorburn, A.W., J.C. Brand, V. Cherikoff, and A.S. Truswell. 1987. Lower postprandial plasma glucose and insulin after addition of *Acacia coriacea* flour to wheat bread. *Aust N Z J Med* 17(1): 24–6.
Tripp, M.L., G. Darland, V.R. Konda, et al. 2012. Optimized mixture of hops rho iso-alpha acids-rich extract and acacia proanthocyanidins-rich extract reduces insulin resistance in 3T3-L1 adipocytes and improves glucose and insulin control in db/db mice. *Nutr Res Pract* 6(5): 405–13.
Vadivel, V., and H.K. Biesalski. 2012. Effect of certain indigenous processing methods on the bioactive compounds of ten different wild type legume grains. *J Food Sci Technol* 49(6): 673–84.
Van Dalen, D. 2020. *Gum Arabic: The Golden Tears of the Acacia Tree*. Leiden, Holland: Leiden University Press.
Wadood, A., N. Wadood, and S.A. Shah. 1989. Effects of *Acacia arabica* and *Caralluma edulis* on blood glucose levels of normal and alloxan diabetic rabbits. *J Pak Med Assoc* 39(8): 208–12.
Wang, P., J.C. Alvarez-Perez, D.P. Felsenfeld, et al. 2015. A high-throughput chemical screen reveals that harmine-mediated inhibition of DYRK1A increases human pancreatic beta cell replication. *Nat Med* 21(4): 383–8.

10 Cosmeceutics
Acacias and Personal Care

As Constantin Hering's famous homeopathic "law of cure" states, problems rise from the depths and exit through the surface. Depths get down, but surfaces are for sharing, and affecting how one is perceived by one's society. People care a lot about this.

Medical students are taught that the *skin*, or dermis/epidermis, is the largest *organ* in the body. It is not really that surprising, but it is possible to be taken aback the first time this concept is encountered. The skin is too easily regarded as a cover, but it is very much the real thing, and it is a real organ. The skin has its own special sense (touch); it keeps water out and lets good air in. It also has its own innervation and touch receptors (Hoffman et al. 2018), and it behaves as a unified endocrine organ that produces hormones, having receptors for and being unitarily affected by growth hormone/insulin-like growth factor-I, neuropeptides, sex steroids, glucocorticoids, retinoids, vitamin D, peroxisome proliferator-activated receptor ligands, eicosanoids, melatonin, and serotonin (Zouboulis 2004).

Acacias play a role in the care of the skin, too. In medicine, the skin is part of the integument, which also includes the hair and nails. In trees, the skin is replaced by the bark, i.e., the pericarp of the stem and its branches.

Gum arabic, i.e., Gum Arabic Senegal, is widely used in the cosmetic industry as an ingredient or additive, and extracts of other acacias are employed as well. In a report with no authors listed published in the *International Journal of Toxicology* in 2005, the safety of several of these species, which were then providing cosmetic ingredients to the world, was considered. The species and plant products employed included: Acacia (*Senegalia*) Catechu Gum, Acacia (*Acacia*) Concinna Fruit Extract, Acacia (*Acacia*) Dealbata Leaf Extract, Acacia Dealbata Leaf Wax, Acacia (*Acacia*) Decurrens Extract (astringent; skin-conditioning agent—occlusive), Acacia (*Vachellia*) Farnesiana Extract (astringent), Acacia Farnesiana Flower Wax, Acacia Farnesiana Gum, and Acacia (*Senegalia*) Senegal Extract. The study, based on a literature survey, concluded that all the uses for Gum Arabic Senegal and its extract were safe; however, for the other ingredients, it was decided that insufficient data were available to reach further conclusions at that time (Anon 2005).

Skin is vulnerable to the environment. To maintain allostatic balance (Sterling 2020), the skin participates proactively in its own defense. Sometimes skin can get caught in the balance, wounded in its line of duty. The two great skin afflictions dispensed by *Acacia s.l.* upon humanity are (1) getting caught by an acacia thorn in unusual places (Collet et al. 2014), and (2) allergy evoked by skin-sticky pollen, as was reported concerning the pollens of *Acacia floribunda* (Ariano et al. 1991), *A. baileyana* (Cvitanović et al. 2004), and other Australian wattles (Howlett et al. 1982). Persons living in close proximity to acacia trees have been found more susceptible to cutaneous diseases caused by *Leishmania donovani* and possibly other *Leishmania* species than those not living close to acacia trees (Elnaiem et al. 2003; Negera et al. 2008), possibly in part due to the role of the sandfly *Phlebotomus orientalis* as the vector associated with *Senegalia seyal* (and *Balanites aegyptiaca*) (Zijlstra and el-Hassan 2001).

This leishmania-transmitting insect inhabits *S. seyal* trees, as well as those of *B. aegyptiaca*. Not only are *S. seyal* involved in the adulteration of gum arabic, but they also give sanctuary to the vector of a pathogen that threatens human life.

In one Australian investigation of native plants used by indigenous people for skin infections, *Acacia falcata* and *A. implexa* were named and experimentally shown to possess antioxidant and antimicrobial properties to support their traditional ethnobotanical usage (Akter et al. 2016). A reference to

DOI: 10.1201/9780429440946-10

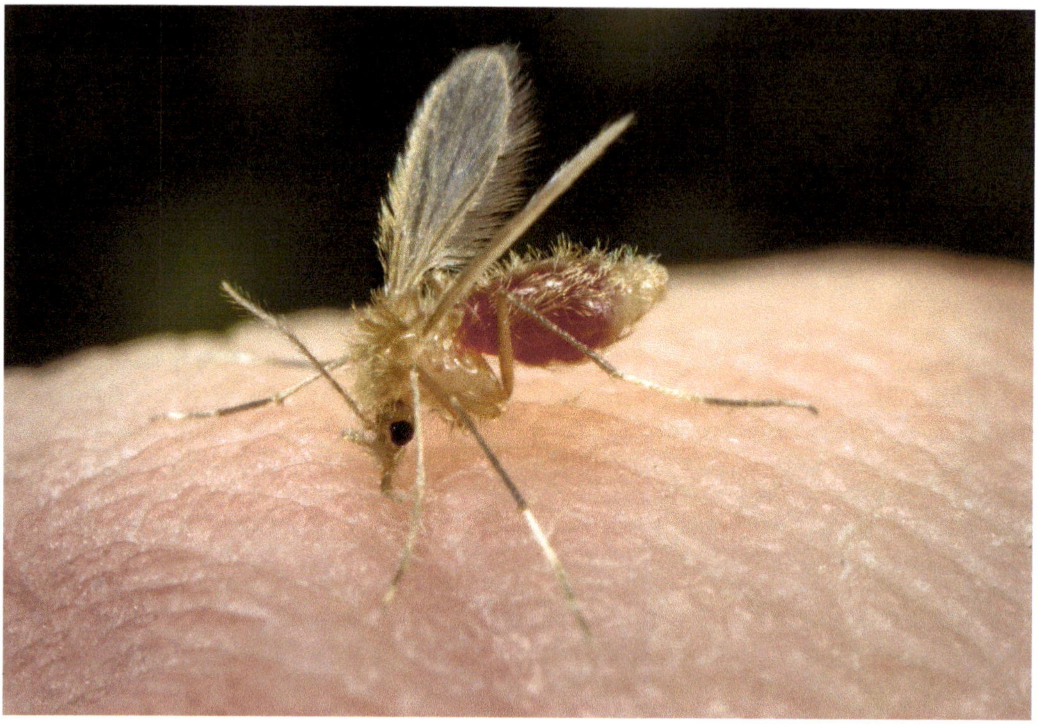

FIGURE 10.1 A *Phlebotomus papatasi* sand fly on the skin surface, sucking blood. (Photo by James Gathany, public domain, via Centers for Disease Control and Prevention, United States Department of Health and Human Services, CDC Public Health Image Library [https://phil.cdc.gov/default.aspx], ID #10275, https://phil.cdc.gov/Details.aspx?pid=10275.)

Vachellia nilotica as *Acacia arabica* was made by an Indian team for its value in the treatment of skin disorders by soothing skin rashes, soreness, inflammation, and burns. They found their "gum acacia" to be effective as a hemostatic, absorptive, and antibiotic, making it a most desirable material for dressing wounds (Bhatnagar et al. 2013). In an *in vitro* study, *Senegalia catechu* extract was combined with extracts of the chestnut tree *Castanea sativa*, the ephedrine-containing Chinese *ma huang* (*Ephedra sinica*), and the high mountain rock exudate *shilajit mumiyo* to create an antibiotic preparation of high safety and efficacy for treatment of acute and chronic skin ulcers (Dashtdar et al. 2013).

In a survey of 87 lay residents (female 78%, male 22%) of a rural South Africa setting where home treatment with medicinal plants constitutes the main form of medical care, *Acacia burkei* was one of the nine plants most frequently named as a therapy for dermatological diseases and disorders (De Wet et al. 2013).

Acacia robusta subsp. *usambarensis* (Taub) Brenan was among the plants most often cited by Tanzanian informants as useful for treating fungal infestations of the skin. An extract of *A. robusta* exhibited strong antifungal activity *in vitro* (Hamza et al. 2006).

Avicins, triterpenoid saponins isolated from *Acacia victoriae*, reduced oxidative and nitrosative stress in mice, preventing skin cancer following exposure to chemical carcinogens (Hanausek et al. 2001). These molecules exert their anti-inflammatory and pro-apoptotic effects in the skin via reducing epidermal hyperplasia, p53 mutations, and *8-hydroxy-2'-deoxyguanosine*, while enhancing apoptosis and the expression of NADPH:quinone oxidoreductase 1 and heme oxygenase-1. The cascade may trigger Nrf2 via "two α,β unsaturated carbonyl groups (Michael reaction sites) in the side chain of the avicin molecule" (Haridas et al. 2004). And because of the avicins' superior skin-penetrating quality, even in extremely small concentrations they facilitate transdermal

FIGURE 10.2 *Acacia floribunda*, "gossamer wattle," is indigenous to New South Wales, Queensland, and Victoria regions of Australia, and may grow to 6 m. It is prized for shade, nitrogen fixation, hedging, and goat fodder. Unsubstantiated reports have indicated that the tree contains dimethyltryptamine. The tree also has a high degree of allergenicity. (a) *A. floribunda* buds, flowers, and foliage. (8/19/2019, Wamban NSW 2537, Australia: by Jenny, CC0 1.0, via iNaturalist, www.inaturalist.org/photos/48649236.) (b) Blossoming *A. floribunda* tree. (8/19/2019, Wamban NSW 2537, Australia: by Jenny, CC0 1.0, via iNaturalist, www.inaturalist.org/photos/48649685.)

FIGURE 10.3 *Acacia baileyana*, "Cootamundra wattle," is indigenous to a very small area of southern New South Wales. The flowers go for commercial decorating in Europe, and the bees like the tree and use it to make excellent honey. (a) *A. baileyana* tree. (2/2/2020, Walkabout Australia, Kangaroo Walk, San Diego Zoo Safari Park, Escondido, CA: by K M, CC BY 2.0, via Flickr, www.flickr.com/photos/131880272@N06/49536514803.) (b) *A. baileyana* flowers. (7/19/2009, South Australia: by Sydney Oats, CC BY 2.0, via Flickr, www.flickr.com/photos/57768042@N00/3733792429.)

(a) (b)

FIGURE 10.4 *Balanites aegyptiaca*, desert date, or Zachum oil tree. (a) *B. aegyptiaca* tree. (1/29/2022, Tinzaouten, Tin Zaoutine, Algeria: by Karim Haddad, CC BY 4.0, via iNaturalist, www.inaturalist.org/photos/178242383.) (b) *B. aegyptiaca* leaves. (5/24/2019, N'Djamena, Tchad: by Denis Bastianelli, CC BY 4.0, via iNaturalist, www.inaturalist.org/photos/112901494.)

penetration of local anesthetics like *lidocaine-HCl*, *prilocaine-HCl*, and *bupivacaine-HCL* through full-thickness porcine skin *ex vivo* (Pino et al. 2014).

Contact dermatitis was caused by allergens isolated and identified from the bark of "Australian blackwood," i.e., *Acacia melanoxylon*, namely *2,6-dimethoxy-1,4-benzoquinone* and *6-methoxy-2-methyl-3,5-dihydrobenzofurano-4,7-dion* ("acamelin," a furanoquinone) (Hausen and Schmalle 1981).

Long-term stable emulsions for cosmeceutical use were created from Solagum AX, a natural polymer blend of acacia and xanthan gums (Hoppel et al. 2014), and a Thai team developed a microemulsion based on an extract of the cosmeceutically valuable "shampoo tree," *Acacia concinna*. An optimized microemulsion incorporating *A. concinna* for cosmeceutical use was achieved with a combination of tea seed *(Camellia oleifera)* oil (5%), polysorbate 85 (40%), ethanol (20%), and water (35%), which exhibited Newtonian flow, 68 nm droplets, and polydispersity index of 0.44 (Poomanee et al. 2017).

A proanthocyanidin-rich extract of *Acacia mearnsii* bark, taken orally over an 8-week period, significantly reduced transepidermal water loss in 33 healthy Japanese adults compared to matched, placebo-taking controls, and also reduced symptoms relating to skin ailments during the 8 weeks (Hoshino et al. 2019). Extract of *A. mearnsii* bark significantly curtailed itching in HR-1 mice when included in their special diet relative to controls, suggesting a putative mechanism of prevention of skin drying via inhibition of atopic dermatitis-associated increased ceramidase expression (Ikarashi et al. 2012).

In 60 post-surgical infants with colostomy tubes, gum acacia (*Senegalia senegal*) was a statistically better protective barrier, resulting in less and milder dermatological inflammation and complications than zinc sulfate matched controls had (Hosseinpour et al. 2012). In an Indian clinical trial, gum arabic was compared to the more costly gum karaya from *Sterculia* spp. trees and to aluminum

FIGURE 10.5 *Camellia oleifera* flower. (2/6/2021, Budock Water, Budock, England: by Tim Green, CC BY 2.0, via Flickr, www.flickr.com/photos/atoach/50915311621.)

paint. Gum arabic was superior to both for healing inflamed, ulcerated, and excoriated skin associated with enterocutaneous fistulae (Kumar et al. 2015). Nanoparticles of ZnO and gum arabic mediated superior and synergistic wound healing effects in rabbits (Manuja et al. 2020), while gum arabic and clove oil were also combined in an anti-inflammatory, analgesic nanoemulsion (Aman et al. 2020). Gum arabic had earlier earned a role in surgical split-skin grafting. The gum arabic was made into a "glue" by gradually adding powdered gum arabic U.S.P. to boiling water until a 50% colloidal solution was obtained. The mucilaginous solution was then sterilized in the autoclave for 15 min at 15 pounds of pressure (Rubin 1945). Allergic sensitivity to gum arabic is known; it can precipitate asthma, which is believed to be mediated by IgE antibodies in response to the polypeptide chains of gum arabic (Fötisch et al. 1998; Sander et al. 2006). In a screen of various polysaccharide complexes including guar gum, xanthan gum, and agar agar on lipid peroxidation within a cream "induced" by UV exposure, gum arabic had a robust dose-dependent protective effect (Trommer and Neubert 2005), and in an earlier study, it similarly protected pancreatic lipase (Child 1945).

An extract of *Vachellia nilotica* was the first cited in a Pakistani-Iraqi-Chinese collaboration studying 28 different medicinal plants for inclusion in anti-aging cosmeceutical creams to protect against ultraviolet radiation. Among the other plants cited were *Benincasa hispida, Calendula officinalis, Camellia sinensis, Nelumbo nucifera, Capparis decidua, Castanea sativa, Coffea arabica, Crocus sativus, Emblica officinalis* Gaertn., *Foeniculum vulgare, Hippophae rhamnoides, Lithospermum erythrorhizon, Malus domestica, Matricaria chamomilla* L., *Moringa oleifera, Morus alba, Ocimum basilicum, Oryza sativa, Polygonum minus, Punica granatum,* and *Silybum marianum* (Jadoon et al. 2015). Also, *V. nilotica* leaf extract was the most potent inhibitor of *7,12-dimethylbenz(a)anthracene (DMBA)*-induced skin carcinogenesis in male Swiss albino mice, followed by extracts of *V. nilotica* flower, followed by extract of *V. nilotica gum* (Meena et al. 2006). Of 73 shrub and tree species belonging to 56 genera and 37 families in a meta-analysis of other studies, *V. nilotica* was the second-most important according to the informant consensus factor, fidelity level, use value, and relative frequency of citation, second in use value only to walnut, *Juglans regia*, and still ahead of *Phyllanthus emblica*, *Pinus roxburghii*, and *Punica granatum* for treating maladies inclusive of skin, hair, and nail disorders (Rashid et al. 2015). The gum arabic tree is a "practically ubiquitous" species worldwide and a source of compounds with antibacterial, antifungal, and antiproliferative properties, including upregulation of the anti-inflammatory cytokine, IL-10 (Sharma et al. 2014).

A catechin-enriched gel was developed to prevent DMBA-induced squamous cell carcinoma in mouse skin *in vivo*, with associated increased activity of glutathione, superoxide dismutase, catalase, glutathione S-transferase, glutathione reductase, and glutathione peroxidase, as well as decreased levels of malondialdehyde (Monga et al. 2014), with a complex *Senegalia catechu* heartwood extract exhibiting a somewhat similar effect (Monga et al. 2011). *S. catechu* is in urgent need of conservation, at least in Nepal, if it is to preserve its status as both an available and valuable medicinal plant for treating dermatological disorders, second only in frequency to gastrointestinal disorders among the medical problems for which Nepalese seek treatment (Singh et al. 2012). The ability of *S. catechu* heartwood extract to modulate inflammatory mediators renders its traditional usage in the treatment of dermatological and other disorders highly justified and worthy of careful heuristic consideration (Sunil et al. 2019).

Ethanolic and ethyl acetate extracts of the South American *Vachellia aroma* had good suppressive activity against skin infestation with both methicillin-sensitive and resistant strains of *Staphylococcus aureus* and *S. epidermidis*, respectively (Mattana et al. 2010). An extract of *Senegalia mellifera* bark were also active against skin-infesting strains of *Staphylococcus aureus*, with such activity also found in pure compounds isolated from the *S. mellifera* extract, namely, the triterpenoids *(20S)-oxolupan-30-al*, *(20R)-oxolupan-30-al*, and *betulinic acid*, with only *(20S)-oxolupan-30-al* active against the mammalian dead skin-infesting dermatophyte (skin fungus) *Microsporum gypseum* (Mutai et al. 2008).

An ethnomedical survey showed *Vachellia gerrardii* to be an important plant used for treating skin afflictions, emphasizing the importance of conservation measures to ensure its adequate supply for future use, as the species is considered endangered (Omwenga et al. 2015).

In general, plant parts of *Acacia s.l.* are among the most popularly used ethnographically for shampoos, soaps, and other indigenous personal care products (Estrada-Castillón et al. 2014). In a desert canyon, male baboons that lived alone liked to eat acacia seeds and figs (Hamilton and Tilson 1982).

And so, the preceding roughly summarizes work done so far in the field of the dermatological phytopharmacology of acacias. However, there is still a largely unexplored frontier that applies very much to *Acacia s.l.*, which relates precisely to the earlier thread that the skin is an organ, even a hormone-responsive and hormone-secreting organ. One of the key pharmacological players at the surface, i.e., on the skin, is serotonin.

Serotonin is famous for its effects on mood and depression due to selective serotonin reuptake inhibitors, and on consciousness, due to its modulation by lysergic acid diethylamide, a partial agonist of 5-HT1A, 5-HT2A, 5-HT2B, 5-HT2C, and 5-HT6 serotonin receptors (Baquiran and Al Khalili 2020) and catalyst in the healing of eczema (Abramson 1976). But serotonin, as is also well known, occurs not only in the brain but throughout the periphery, in, but certainly not restricted to, the peripheral nervous system. Serotonin is mainly stored in the body in platelets, and via these platelets and with the help of serotonin "rickshaws," i.e., serotonin transporter proteins, it can, so to speak, intra-somatically get around.

Of course, serotonin and serotonin receptors are no strangers to the skin, but they curiously appear when there is trouble. Serotonin is probably best known as a dermatic irritant, causing itch via central neural circuits (Cevikbas et al. 2011) as well as in the periphery (Hägermark 1992), including crosstalk with histaminic processes and mast cells (Kritas et al. 2014), and also triggering pain and transmitting signals of pain through central serotonergic signaling (Eide and Hole 1991; Liu et al. 2007), including via serotonin receptor proteins 5-HT1A and 5-HT 1B (Eide et al. 1990), and 5-HT3A, via 5-HT transporter protein (Kayser et al. 2007), and in a craniofacial setting, 5-HT2A (Okamoto et al. 2007). Serotonin also activates human monocytes and prevents apoptosis, allowing (cutaneous) inflammation to continue (Soga et al. 2007).

Specifically, serotonin receptors on the skin, like those everywhere else, are putatively *tuned* in their responsiveness to serotonin, according to the actual or perceived requirements of the body in different states of health and disease. So, psoriasis, whose name stems from the Greek word for "itch," is a disturbance that can produce itch and be excoriating, disfiguring, and intensely dysphoric. It is serotonergically tuned, with new serotonin receptors and fresh serotonin to "continue operations" for the disease (Nordlind et al. 2006), thus raising interest in serotonergic drugs to treat psoriasis (Thorslund and Nordlind 2007). Itch itself, for example, cholestatic itch, is mediated by both central and peripheral mechanisms (Tian et al. 2016), in which, in both cases, serotonin appears to play a commanding role (Yamaguchi et al. 1999). Similarly, the same mechanism of making new serotonin receptors and releasing free serotonin applies for both allergic contact-induced (El-Nour et al. 2007a, 2007b; Lundeberg et al. 1999, 2002; Wetterberg et al. 2011) and stress-induced atopic dermatitides (Lonne-Rahm et al. 2008; Rasul et al. 2011, 2013, 2016), dermatologic cancer (Naimi-Akbar et al. 2010); pigmentation of skin cells in response to stress in mice (Wu et al. 2014), and mastocytosis (Ritter et al. 2012), which may be cutaneous. Serotonin via the 5-HT2A receptors is also associated with melanogenesis, with great relevance to future dermatologic health and disease (Lee et al. 2011), and with trans-urocanic acid-induced immunological suppression following exposure to ultraviolet radiation (Walterscheid et al. 2006). Further, a deeper understanding of the serotonergic changes occurring during itch has been elaborated (Akiyama et al. 2016). Zabolinejad et al. (2019) showed elevated serotonin transport protein (SERT) expression in people with chronic spontaneous urticaria, suggesting an associative if not causative role for SERT in the progression of the disease.

If the skin is truly an organ with its own innervation network (Blessing and Seaman 2003), tactile receptors, respiration, and allosteric responsivity to circumstance, then perhaps all of that can be modulated selectively through the skin's serotonin receptors. Perhaps there is a "mind" within those cells (Satprem 1981) that can be therapeutically reached with the right phytoceutical tuning agent. This is not a new idea. Psychoactive drugs that affect serotonin, like antidepressants, have been tried or suggested for skin problems (Bojanovská 1968; Tennyson and Levine 2001).

Psychedelics, too, are not new, and are found in natural sources: plants, mushrooms, and animals. Pertinent to the theme of this chapter, the animal sources include the secreted psychedelic compounds that occur in the *skin* of amphibians, most famously toads and frogs. For example, it has been suggested that the story of the toad and the princess is based on the fact that the toad contains the hallucinogenic indole alkaloid bufotenine, found in and named for the skin of common toads (*Bufo bufo*). She licks the toad, ingests the compound, undergoes psychic upheaval, and transforms (Siegel and McDaniel 1991). Dimethyltryptamine (DMT), 5-methoxy DMT, and methyltryptamine, such as found in acacia species, are also those found in the *skin* of toads and other amphibians (Erspamer et al. 1966; Zulfiker et al. 2016), and they have greater psychedelic potency and considerably better safety peripherally than bufotenine. Other related indole alkaloids in toad skin have been noted, such as the O-sulphate of bufotenidine and *2-(3-indolyl) ethyltrimethylammonium*, a quaternary ammonium base of tryptamine (Roseghini et al. 1976), and their metabolism *in situ* has been elaborated (Sakai et al. 1965).

Could putative "tuning agents," such as harmine, affect peripheral cells, including skin cells, especially when combined with dimethyltryptamines found in parts of certain *Acacia s.l.* species? Could the compounds classically known to "reset" the psyche also "reset" the skin? Could *Acacia s.l.* help to provoke a deeply therapeutic "dermatodelic" effect through selective modulation and positive "re-tuning" of dermatic serotonergic receptors and their activity, homologous to the aforementioned lipidelic effect in Chapter 8 of this book on obesity, the somatodelic effect in lymphocytes (Lansky et al. 2017), and the ocudelic effect in the eye (Lansky 2020b), all putatively initiated by naturally occurring indolic hallucinogens?

To test such an hypothesis, gels could be compounded, certainly with acacia gum, but also with extracts of *Acacia s.l.* parts containing known amounts of substituted tryptamines such as *N-methyltryptamine*, *N,N-dimethyltryptamine*, and *5-methoxy-dimethyltryptamine*. The acacia extracts could be combined with an extract of a plant such as *Peganum harmala* (Lansky et al. 2017), which is rich in reversible monoamine oxidase inhibitory harmala alkaloids that can help keep the DMT and congeners from being oxidized and also shows its own potent protective benefit against dermatological diseases (El-Saad and El-Rifaie 1980).

It is known that set (mind-set) and setting (treatment space) are of great importance in conventional psychedelic psychotherapy, and a protocol for a method originally for LSD therapy employing both a male and a female therapist, high fidelity, recorded, stage-specific classical music, eyeshades, and headphones in a cozy home-like treatment (inner travel) room has been described in detail (Bonny and Pahnke 1972). The methods summarized in that paper set the standard for all or most medical work with psychedelics in the United States that followed. For testing purposes, the method could be adopted in dermatology, with dermatological assessments made before treatment and at different points after it.

Similar studies could also be done in animals *in vivo* or in skin organ cultures, with external conditions, such as light and music, similarly controlled. Oral dosing, as was done with other acacia phytoceuticals from *A. mearnsii*, could be similarly tested with alkaloid-rich fractions *per orum* of extracts of selected *Acacia s.l.* species, using the opportunity to test various dermatological and inflammatory parameters and in clinical scenarios to review and attend to

FIGURE 10.6 *Acacia confusa* tree trunk. The bark of *A. confusa* contains potentially psychoactive DMTs. (8/12/2004, Kauhikoa Hill, Maui, Hawaii: by Forest & Kim Starr, CC BY 4.0, via Starr Environmental, www.starrenvironmental.com/images/image/?q=24419769450.)

psychodynamic issues as well as subjective improvement in itching, pain, or other dermatological dysphoria. As a specific sigma-1 receptor agonist, DMT contained in acacias may benefit many diseases, including psoriasis, via mechanisms that bypass serotonin receptors (Bourrie et al. 2004). Also, psoriasis was one of the diseases reported as being at least temporarily resolved for long periods following treatment with ayahuasca during the Shipibo ceremony in Peru, in which a dimethyltryptamine-containing plant such as *Psychotria viridis* is included in the aqueous extraction along with the harmala alkaloid-rich *Banisteriopsis caapi.* The therapeutic change may have occurred not only on the mental and emotional level, i.e., in the mind, but also at a more fundamental physical level, perhaps related to epigenetic modifications (Tafur 2017), without the aegis of the central nervous system, via serotonin receptors in the periphery, including the skin (Lansky 2020a, b).

Also of note is that a derivate of serotonin, *melatonin*, is also found in the skin, and it may cyclize under certain circumstances to yield a potent endogenous hallucinogen, *10-methoxy harmalan,* which is also a harmala alkaloid (Slominski et al. 2018), thus "closing a circle."

REFERENCES

Abramson, H.A. 1976. Reassociation of dreams. II. An LSD study of sexual conflicts in eczema and asthma. *J Asthma Res* 13(4): 193–233.

Akiyama, T., M. Ivanov, M. Nagamine, et al. 2016. Involvement of TRPV4 in serotonin-evoked scratching. *J Invest Dermatol* 136(1): 154–60.

Akter, K., E.C. Barnes, J.J. Brophy, et al. 2016. Phytochemical profile and antibacterial and antioxidant activities of medicinal plants used by Aboriginal people of New South Wales, Australia. *Evid Based Complement Alternat Med* 2016: 4683059.

Aman, R.M., I.I. Abu Hashim, and M.M. Meshali. 2020. Novel clove essential oil nanoemulgel tailored by Taguchi's model and scaffold-based nanofibers: Phytopharmaceuticals with promising potential as cyclooxygenase-2 inhibitors in external inflammation. *Int J Nanomedicine* 15: 2171–95.

Anon. 2005. Final report of the safety assessment of *Acacia catechu* gum, *Acacia concinna* fruit extract, *Acacia dealbata* leaf extract, *Acacia dealbata* leaf wax, *Acacia decurrens* extract, *Acacia farnesiana* extract, *Acacia farnesiana* flower wax, *Acacia farnesiana* gum, *Acacia senegal* extract, *Acacia senegal* gum, and *Acacia senegal* gum extract. *Int J Toxicol* 24(Suppl 3): 75–118.

Ariano, R., R.C. Panzani, and J. Amedeo. 1991. Pollen allergy to mimosa (*Acacia floribunda*) in a Mediterranean area: An occupational disease. *Ann Allergy* 66(3): 253–6.

Baquiran, M., and Y. Al Khalili. 2020. Lysergic acid diethylamide (LSD, Entactogen) toxicity. In: *StatPearls* [Internet]. Treasure Island, FL: StatPearls Publishing. www.ncbi.nlm.nih.gov/books/NBK553216/.

Bhatnagar, M., L. Parwani, V. Sharma, J. Ganguli, and A. Bhatnagar. 2013. Hemostatic, antibacterial biopolymers from *Acacia arabica* (Lam.) Willd. and *Moringa oleifera* (Lam.) as potential wound dressing materials. *Indian J Exp Biol* 51(10): 804–10.

Blessing, W.W., and B. Seaman. 2003. 5-Hydroxytryptamine(2A) receptors regulate sympathetic nerves constricting the cutaneous vascular bed in rabbits and rats. *Neuroscience* 117(4): 939–48.

Bojanovská, A. 1968. Psychofarmaká a ich mozné pouzitie v dermatológii [Psychopharmaca and their possible use in dermatology]. *Cesk Dermatol* 43(5): 338–42.

Bonny, H.L., and W.N. Pahnke. 1972. The use of music in psychedelic (LSD) psychotherapy. *J Music Therapy* 9(2): 64–87.

Bourrie, B., E. Bribes, J.M. Derocq, H. Vidal, and P. Casellas. 2004. Sigma receptor ligands: Applications in inflammation and oncology. *Curr Opin Investig Drugs* 5(11): 1158–63.

Cevikbas, F., M. Steinhoff, and A. Ikoma. 2011. Role of spinal neurotransmitter receptors in itch: New insights into therapies and drug development. *CNS Neurosci Ther* 17(6): 742–9.

Child, G.P. 1945. The protection of pancreatic lipase by acacia gum. *Science* 101(2621): 311–12.

Collet, C., N. Dumont, M. Romeu, N. Graillon, and L. Guyot. 2014. Une lésion dermatologique paranasale atypique chez un enfant [An unusual paranasal dermatologic lesion in a child. Acacia thorn]. *Rev Stomatol Chir Maxillofac Chir Orale* 115(1): e9–e11.

Cvitanović, S., L. Znaor, D. Perisić, and D. Grbić. 2004. Hypersensitivity to pollen allergens on the Adriatic coast. *Arh Hig Rada Toksikol* 55(2–3): 147–54.

Dashtdar, M., M.R. Dashtdar, B. Dashtdar, M.K. Shirazi, and S.A. Khan. 2013. In-vitro, anti-bacterial activities of aqueous extracts of *Acacia catechu* (L.F.) Willd, *Castanea sativa*, *Ephedra sinica* Stapf and shilajita mumiyo against gram positive and gram negative bacteria. *J Pharmacopuncture* 16(2): 15–22.

De Wet, H., S. Nciki, and S.F. van Vuuren. 2013. Medicinal plants used for the treatment of various skin disorders by a rural community in northern Maputaland, South Africa. *J Ethnobiol Ethnomed* 9: 51.

Eide, P.K., and K. Hole. 1991. Interactions between serotonin and substance P in the spinal regulation of nociception. *Brain Res* 550(2): 225–30.

Eide, P.K., N.M. Joly, and K. Hole. 1990. The role of spinal cord 5-HT1A and 5-HT1B receptors in the modulation of a spinal nociceptive reflex. *Brain Res* 536(1–2): 195–200.

Elnaiem, D.E., A.M. Mukhawi, M.M. Hassan, et al. 2003. Factors affecting variations in exposure to infections by *Leishmania donovani* in eastern Sudan. *East Mediterr Health J* 9(4): 827–36.

El-Nour, H., L. Lundeberg, N. Abdel-Magid, S.B. Lonne-Rahm, E.C. Azmitia, and K. Nordlind. 2007a. Serotonergic mechanisms in human allergic contact dermatitis. *Acta Derm Venereol* 87(5): 390–6.

El-Nour, H., L. Lundeberg, A. Boman, D. Abramowski, M. Holst, and K. Nordlind. 2007b. The expression and functional significance of the serotonin(2C) receptor in murine contact allergy. *Exp Dermatol* 16(8): 644–50.

El-Saad El-Rifaie, M. 1980. *Peganum harmala*: Its use in certain dermatoses. *Int J Dermatol* 19(4): 221–2.

Erspamer, V., M. Roseghini, R. Endean, and A. Anastasi. 1966. Biogenic amines and active polypeptides in the skin of Australian amphibians. *Nature* 212(5058): 204.

Estrada-Castillón, E., M. Garza-López, J.Á. Villarreal-Quintanilla, et al. 2014. Ethnobotany in Rayones, Nuevo León, México. *J Ethnobiol Ethnomed* 10: 62.

Fötisch, K., J. Fäh, B. Wüthrich, F. Altmann, D. Haustein, and S. Vieths. 1998. IgE antibodies specific for carbohydrates in a patient allergic to gum arabic (*Acacia senegal*). *Allergy* 53(11): 1043–51.

Hägermark, O. 1992. Peripheral and central mediators of itch. *Skin Pharmacol* 5(1): 1–8.

Hamilton, W.J., 3rd, and R.L. Tilson. 1982. Solitary male chacma baboons in a desert canyon. *Am J Primatol* 2(2): 149–58.

Hamza, O.J., C.J. van den Bout-van den Beukel, M.I. Matee, et al. 2006. Antifungal activity of some Tanzanian plants used traditionally for the treatment of fungal infections. *J Ethnopharmacol* 108(1): 124–32.

Hanausek, M., P. Ganesh, Z. Walaszek, C.J. Arntzen, T.J. Slaga, and J.U. Gutterman. 2001. Avicins, a family of triterpenoid saponins from *Acacia victoriae* (Bentham), suppress H-ras mutations and aneuploidy in a murine skin carcinogenesis model. *Proc Natl Acad Sci U S A* 98(20): 11551–6.

Haridas, V., M. Hanausek, G. Nishimura, et al. 2004. Triterpenoid electrophiles (avicins) activate the innate stress response by redox regulation of a gene battery. *J Clin Invest* 113(1): 65–73.

Hausen, B.M., and H. Schmalle. 1981. Quinonoid constituents as contact sensitisers in Australian blackwood (*Acacia melanoxylon* RBR). *Br J Ind Med* 38(2): 105–9.

Hoffman, B.U., Y. Baba, T.N. Griffith, et al. 2018. Merkel cells activate sensory neural pathways through adrenergic synapses. *Neuron* 100(6): 1401–13.e6.

Hoppel, M., D. Mahrhauser, C. Stallinger, F. Wagner, M. Wirth, and C. Valenta. 2014. Natural polymer-stabilized multiple water-in-oil-in-water emulsions: A novel dermal drug delivery system for 5-fluorouracil. *J Pharm Pharmacol* 66(5): 658–67.

Hoshino, T., S.I. Yamashita, N. Suzuki, A. Baba, S. Ogawa, and T. Izumi. 2019. Impact of acacia bark extract tablets on the skin of healthy humans: A randomized, double-blind, placebo-controlled study. *Biosci Biotechnol Biochem* 83(3): 538–50.

Hosseinpour, M., A. Fazeli, and M. Agabeigi. 2012. Efficacy of *Acacia senegal* for stoma care in children with colostomy. *Eur J Pediatr Surg* 22(3): 234–7.

Howlett, B.J., D.J. Hill, and R.B. Knox. 1982. Cross-reactivity between *Acacia* (wattle) and rye grass pollen allergens. Detection of allergens in *Acacia* (wattle) pollen. *Clin Allergy* 12(3): 259–68.

Ikarashi, N., W. Sato, T. Toda, M. Ishii, W. Ochiai, and K. Sugiyama. 2012. Inhibitory effect of polyphenol-rich fraction from the bark of *Acacia mearnsii* on itching associated with allergic dermatitis. *Evid Based Complement Alternat Med* 2012: 120389.

Jadoon, S., S. Karim, M.H. Bin Asad, et al. 2015. Anti-aging potential of phytoextract loaded-pharmaceutical creams for human skin cell longevity. *Oxid Med Cell Longev* 2015: 709628.

Kayser, V., I.E. Elfassi, B. Aubel, et al. 2007. Mechanical, thermal and formalin-induced nociception is differentially altered in 5-HT1A-/-, 5-HT1B-/-, 5-HT2A-/-, 5-HT3A-/- and 5-HTT-/- knock-out male mice. *Pain* 130(3): 235–48.

Kritas, S.K., A. Saggini, G. Cerulli, et al. 2014. Relationship between serotonin and mast cells: Inhibitory effect of anti-serotonin. *J Biol Regul Homeost Agents* 28(3): 377–80.

Kumar, P., and Ahmad S. Namrata. 2015. Enterocutaneous fistula: Different surgical intervention techniques for closure along with comparative evaluation of aluminum paint, karaya gum (Hollister) and gum acacia for peristomal skin care. *J Clin Diagn Res* 9(12): PC16–PC20.

Lansky, E.S. 2020a. [Remembering the self: A review of Dr Joseph Tafur's *The fellowship of the river.*] [In Hebrew.] *Harefuah* 159(1): 55–7.

Lansky, E.S. 2020b. Novel harmala-ocudelic tuning (HOT) for ocular disorders. *Med Hypotheses* 143: 109834.

Lansky, E.S., H.M. Paavilainen, and S. Lansky. 2017. *Harmal: The Genus Peganum.* Boca Raton, FL: CRC Press.

Lee, H.J., M.K. Park, S.Y. Kim, H.Y. Park Choo, A.Y. Lee, and C.H. Lee. 2011. Serotonin induces melanogenesis via serotonin receptor 2A. *Br J Dermatol* 165(6): 1344–8.

Liu, F.Y., G.G. Xing, X.X. Qu, I.S. Xu, J.S. Han, and Y. Wan. 2007. Roles of 5-hydroxytryptamine (5-HT) receptor subtypes in the inhibitory effects of 5-HT on C-fiber responses of spinal wide dynamic range neurons in rats. *J Pharmacol Exp Ther* 321(3): 1046–53.

Lonne-Rahm, S.B., H. Rickberg, H. El-Nour, P. Mårin, E.C. Azmitia, and K. Nordlind. 2008. Neuroimmune mechanisms in patients with atopic dermatitis during chronic stress. *J Eur Acad Dermatol Venereol* 22(1): 11–18.

Lundeberg, L., H. El-Nour, S. Mohabbati, M. Morales, E. Azmitia, and K. Nordlind. 2002. Expression of serotonin receptors in allergic contact eczematous human skin. *Arch Dermatol Res* 294(9): 393–8.

Lundeberg, L., V. Mutt, and K. Nordlind. 1999. Inhibitory effect of vasoactive intestinal peptide on the challenge phase of allergic contact dermatitis in humans. *Acta Derm Venereol* 79(3): 178–82.

Manuja, A., R. Raguvaran, B. Kumar, A. Kalia, and B.N. Tripathi. 2020. Accelerated healing of full thickness excised skin wound in rabbits using single application of alginate/acacia based nanocomposites of ZnO nanoparticles. *Int J Biol Macromol* 155: 823–33.

Mattana, C.M., S.E. Satorres, A. Sosa, M. Fusco, and L.E. Alcará. 2010. Antibacterial activity of extracts of *Acacia aroma* against methicillin-resistant and methicillin-sensitive *Staphylococcus. Braz J Microbiol* 41(3): 581–7.

Meena, P.D., P. Kaushik, S. Shukla, A.K. Soni, M. Kumar, and A. Kumar. 2006. Anticancer and antimutagenic properties of *Acacia nilotica* (Linn.) on 7,12-dimethylbenz(a)anthracene-induced skin papillomagenesis in Swiss albino mice. *Asian Pac J Cancer Prev* 7(4): 627–32.

Monga, J., V. Aggarwal, S.K. Suthar, M. Chauhan, K. Nongalleima, and M. Sharma. 2014. Topical (+)-catechin emulsified gel prevents DMBA/TPA-induced squamous cell carcinoma of the skin by modulating antioxidants and inflammatory biomarkers in Balb/c mice. *Food Funct* 5(12): 3197–3207.

Monga, J., C.S. Chauhan, and M. Sharma. 2011. Human epithelial carcinoma cytotoxicity and inhibition of DMBA/TPA induced squamous cell carcinoma in Balb/c mice by *Acacia catechu* heartwood. *J Pharm Pharmacol* 63(11): 1470–82.

Mutai, C., C. Bii, G. Rukunga, et al. 2008. Antimicrobial activity of pentacyclic triterpenes isolated from *Acacia mellifera. Afr J Tradit Complement Altern Med* 6(1): 42–8.

Naimi-Akbar, C., M. Ritter, S. Demel, et al. 2010. Different serotonergic expression in nevomelanocytic tumors. *Cancers (Basel)* 2(2): 1166–77.

Negera, E., E. Gadisa, L. Yamuah, et al. 2008. Outbreak of cutaneous leishmaniasis in Silti woreda, Ethiopia: Risk factor assessment and causative agent identification. *Trans R Soc Trop Med Hyg* 102(9): 883–90.

Nordlind, K., K. Thorslund, S. Lonne-Rahm, et al. 2006. Expression of serotonergic receptors in psoriatic skin. *Arch Dermatol Res* 298(3): 99–106.

Okamoto, K., H. Imbe, A. Kimura, T. Donishi, Y. Tamai, and E. Senba. 2007. Activation of central 5HT2A receptors reduces the craniofacial nociception of rats. *Neuroscience* 147(4): 1090–102.

Omwenga, E.O., A. Hensel, A. Shitandi, and F.M. Goycoolea. 2015. Ethnobotanical survey of traditionally used medicinal plants for infections of skin, gastrointestinal tract, urinary tract and the oral cavity in Borabu Sub-County, Nyamira County, Kenya. *J Ethnopharmacol* 176: 508–14.

Pino, C.J., M.A. Scherer, and V.P. Shastri. 2014. Investigation of the transdermal transport of charged local anesthetics in the presence of triterpene saponin glycosides. *Drug Deliv Transl Res* 4(2): 131–8.

Poomanee, W., W. Chaiyana, R.R. Wickett, and P. Leelapornpisid. 2017. Stability and solubility improvement of Sompoi (*Acacia concinna* Linn.) pod extract by topical microemulsion. *Asian J Pharm Sci* 12(4): 386–93.

Rashid, S., M. Ahmad, M. Zafar, et al. 2015. Ethnobotanical survey of medicinally important shrubs and trees of Himalayan region of Azad Jammu and Kashmir, Pakistan. *J Ethnopharmacol* 166: 340–51.

Rasul, A., H. El-Nour, R.D. Blakely, et al. 2011. Effect of chronic mild stress on serotonergic markers in the skin and brain of the NC/Nga atopic-like mouse strain. *Arch Dermatol Res* 303(9): 625–33.

Rasul, A., H. El-Nour, S.B. Lonne-Rahm, et al. 2016. Serotonergic markers in atopic dermatitis. *Acta Derm Venereol* 96(6): 732–6.

Rasul, A., K. Nordlind, and C.F. Wahlgren. 2013. Pruritic and vascular responses induced by serotonin in patients with atopic dermatitis and in healthy controls. *Acta Derm Venereol* 93(3): 277–80.

Ritter, M., H. El-Nour, M.A. Hedblad, et al. 2012. Serotonin and its 5-HT1 receptor in human mastocytosis. *Immunopharmacol Immunotoxicol* 34(4): 679–85.

Roseghini, M., R. Endean, and A. Temperilli. 1976. New and uncommon indole- and imidazole-alkylamines in skins of amphibians from Australia and Papua New Guinea. *Z Naturforsch C Biosci* 31(3–4): 118–20.

Rubin, L.R. 1945. Simplification of split skin grafting; gum acacia technic. *Am J Surg* 70: 302–7.

Sakai, K., M. Fuse, and K. Nakai. 1965. Indolealkylamine metabolism in the toad. *Jpn J Pharmacol* 15(4): 443–4.

Sander, I., M. Raulf-Heimsoth, K. Wiemer, S. Kespohl, T. Brüning, and R. Merget. 2006. Sensitization due to gum arabic (*Acacia senegal*): The cause of occupational allergic asthma or crossreaction to carbohydrates? *Int Arch Allergy Immunol* 141(1): 51–6.

Satprem. 1981. *The Mind of the Cells, or Willed Mutation of Our Species*. Pondicherry: Institute for Evolutionary Research.

Sharma, A.K., A. Kumar, S.K. Yadav, and A. Rahal. 2014. Studies on antimicrobial and immunomodulatory effects of hot aqueous extract of *Acacia nilotica* L. Leaves against common veterinary pathogens. *Vet Med Int* 2014: 747042.

Siegel, D.M., and S.H. McDaniel. 1991. The frog prince: Tale and toxicology. *Am J Orthopsychiatry* 61(4): 558–62.

Singh, A.G., A. Kumar, and D.D. Tewari. 2012. An ethnobotanical survey of medicinal plants used in Terai forest of western Nepal. *J Ethnobiol Ethnomed* 8: 19.

Slominski, A.T., R. Hardeland, M.A. Zmijewski, R.M. Slominski, R.J. Reiter, and R. Paus. 2018. Melatonin: A cutaneous perspective on its production, metabolism, and functions. *J Invest Dermatol* 138(3): 490–9.

Soga, F., N. Katoh, T. Inoue, and S. Kishimoto. 2007. Serotonin activates human monocytes and prevents apoptosis. *J Invest Dermatol* 127(8): 1947–55.

Sterling, P. 2020. *What Is Health? Allostasis and the Evolution of Human Design*. Cambridge, MA: MIT Press.

Sunil, M.A., V.S. Sunitha, E.K. Radhakrishnan, and M. Jyothis. 2019. Immunomodulatory activities of *Acacia catechu*, a traditional thirst quencher of South India. *J Ayurveda Integr Med* 10(3): 185–91.

Tafur, J. 2017. *The Fellowship of the River: A Medical Doctor's Exploration into Traditional Amazonian Plant Medicine*. Phoeniz, AZ: Espiritu Books.

Tennyson, H., and N. Levine. 2001. Neurotropic and psychotropic drugs in dermatology. *Dermatol Clin* 19(1): 179–97, x.

Thorslund, K., and K. Nordlind. 2007. Serotonergic drugs—A possible role in the treatment of psoriasis? *Drug News Perspect* 20(8): 521–5.

Tian, B., X.L. Wang, Y. Huang, et al. 2016. Peripheral and spinal 5-HT receptors participate in cholestatic itch and antinociception induced by bile duct ligation in rats. *Sci Rep* 6: 36286.

Trommer, H., and R.H. Neubert. 2005. The examination of polysaccharides as potential antioxidative compounds for topical administration using a lipid model system. *Int J Pharm* 298(1): 153–63.

Walterscheid, J.P., D.X. Nghiem, N. Kazimi, et al. 2006. Cis-urocanic acid, a sunlight-induced immunosuppressive factor, activates immune suppression via the 5-HT2A receptor. *Proc Natl Acad Sci U S A* 103(46): 17420–5.

Wetterberg, J., C. Taher, E.C. Azmitia, and H. El-Nour. 2011. Time-dependent modulation of serotonin and its receptors 1A and 2A expression in allergic contact dermatitis. *J Eur Acad Dermatol Venereol* 25(10): 1200–5.

Wu, H.L., S.L. Pang, Q.Z. Liu, Q. Wang, M.X. Cai, and J. Shang. 2014. 5-HT1A/1B receptors as targets for optimizing pigmentary responses in C57BL/6 mouse skin to stress. *PLoS One* 9(2): e89663.

Yamaguchi, T., T. Nagasawa, M. Satoh, and Y. Kuraishi. 1999. Itch-associated response induced by intradermal serotonin through 5-HT2 receptors in mice. *Neurosci Res* 35(2): 77–83.

Zabolinejad, N., S. Molkara, B. Bakhshodeh, H. Ghaffari-Nazari, and M. Khoshkhui. 2019. The expression of serotonin transporter protein in the skin of patients with chronic spontaneous urticaria and its relation with depression and anxiety. *Arch Dermatol Res* 311(10): 825–31.

Zijlstra, E.E., and A.M. el-Hassan. 2001. Leishmaniasis in Sudan. Visceral leishmaniasis. *Trans R Soc Trop Med Hyg* 95(Suppl 1): S27–S58.

Zouboulis, C.C. 2004. The human skin as a hormone target and an endocrine gland. *Hormones (Athens)* 3(1): 9–26.

Zulfiker, A.H., G.L. Mariottini, J. Qi, I.D. Grice, and M.Q. Wei. 2016. Indolealkylamines from toad vertebrates and sea invertebrates—Their identification and potential activities on the central nervous system. *Cent Nerv Syst Agents Med Chem* 16(3): 197–207.

Appendix: Structures of Chemical Compounds

The main chemical compounds mentioned in the text are presented here in their approximate order of appearance. The reader is also referred to the Index.

Chapter		Name	Structure
1	**Structure 1**	N-Methyltryptamine	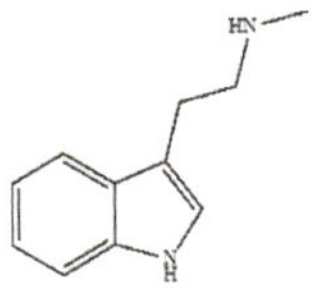
	Structure 2	N,N-Dimethyltryptamine	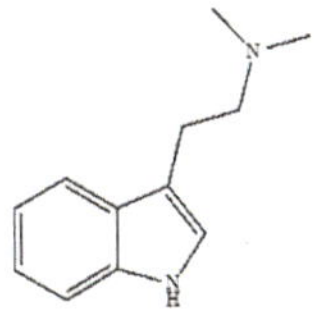
2	**Structure 3**	Catechol	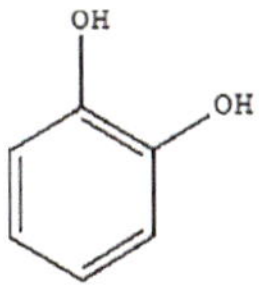
	Structure 4	Tyramine	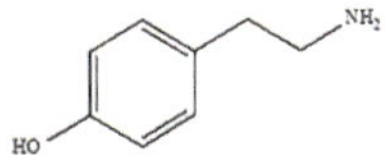
	Structure 5	Phenethylamine	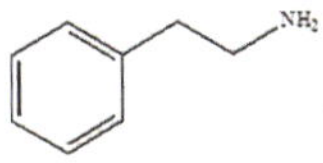
	Structure 6	Amphetamine	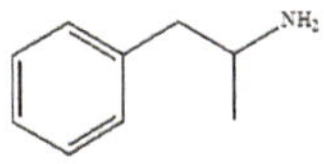
	Structure 7	Nicotine	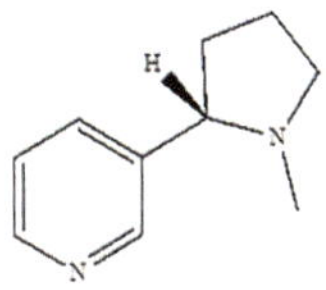
	Structure 8	Mescaline	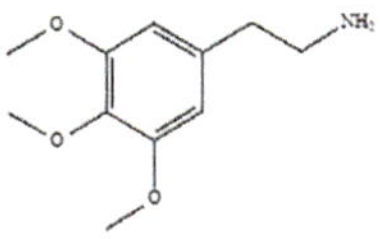

Chapter	Name	Structure
2 (*Continued*)	**Structure 9** Epicatechin	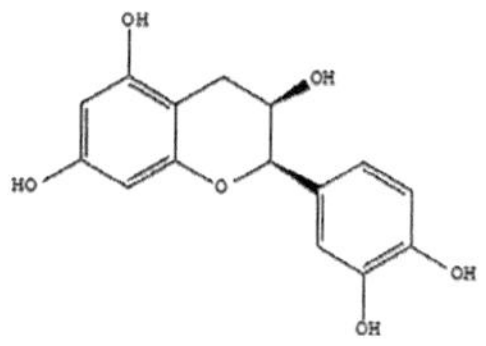
	Structure 10 Catechin	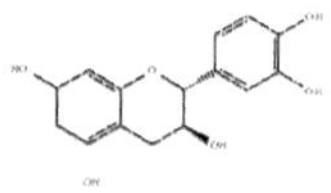
	Structure 11 3,5-Dihydroxy-4-methoxy-benzoic acid (4-OMGA)	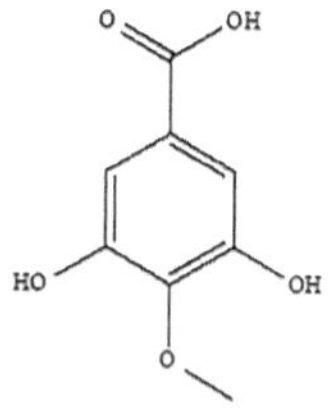
	Structure 12 Gallic acid	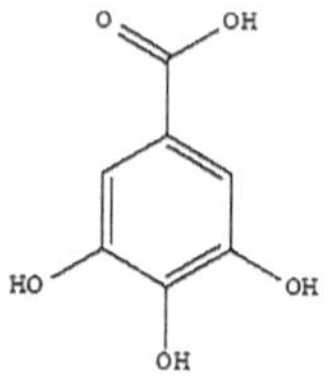
	Structure 13 Methyl gallate	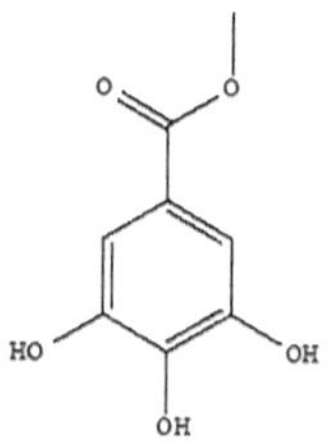
	Structure 14 Ethyl gallate	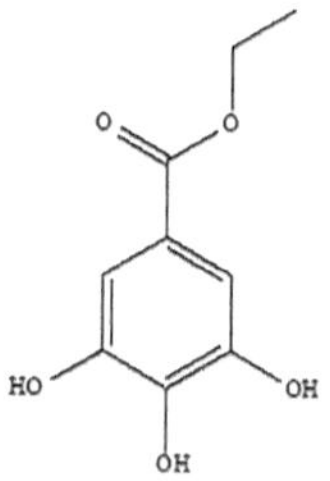
	Structure 15 Pyrogallol	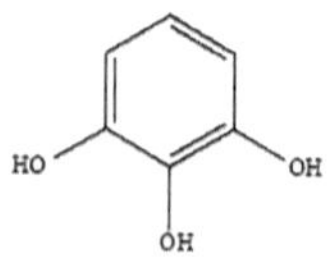
	Structure 16 Niloticane ((3β,12α)-3,12-dihydroxy-14-methyl-13-vinylpodocarp-13-en-11-one)	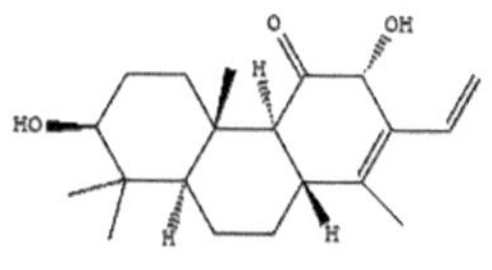

Chapter		Name	Structure
2 *(Continued)*	**Structure 17**	Leucofisetinidin	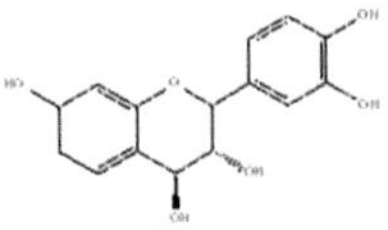
	Structure 18	Acaciaside a (Adapted with permission from Mahato, S.B., B.C. Pal, and A.K. Nandy. 1992. Structure elucidation of two acylated triterpenoid bisglycosides from *Acacia auriculiformis* Cunn. *Tetrahedron* 48(32): 6717–28. https://doi.org/10.1016/S0040-4020(01)80017-X.)	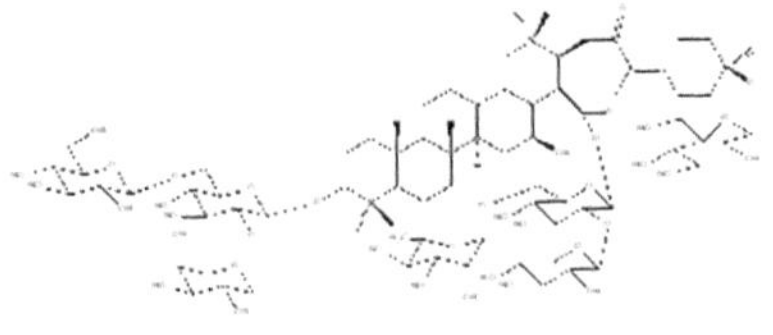
	Structure 19	Acaciaside b (Adapted with permission from Mahato, S.B., B.C. Pal, and A.K. Nandy. 1992. Structure elucidation of two acylated triterpenoid bisglycosides from *Acacia auriculiformis* Cunn. *Tetrahedron* 48(32): 6717–28. https://doi.org/10.1016/S0040-4020(01)80017-X.)	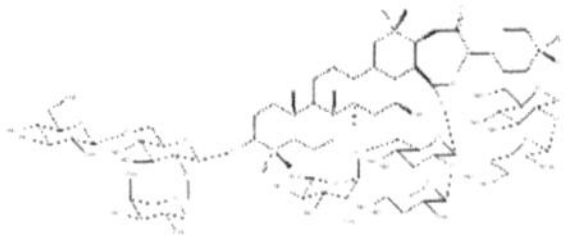
	Structure 20	Tetracosanoic acid (2S)-2,3-dihydroxypropyl ester	
	Structure 21	(3β,22E)-Stigmasta-5,22-dien-3-yl β-D-glucopyranoside	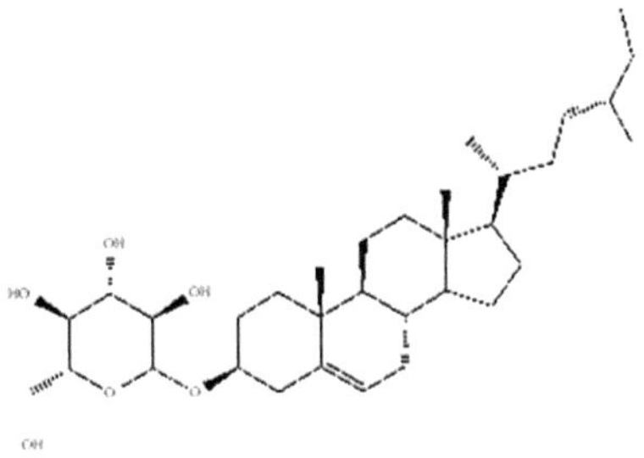
	Structure 22	Prunin ((2S) naringenin 7-O-β-glucopyranoside)	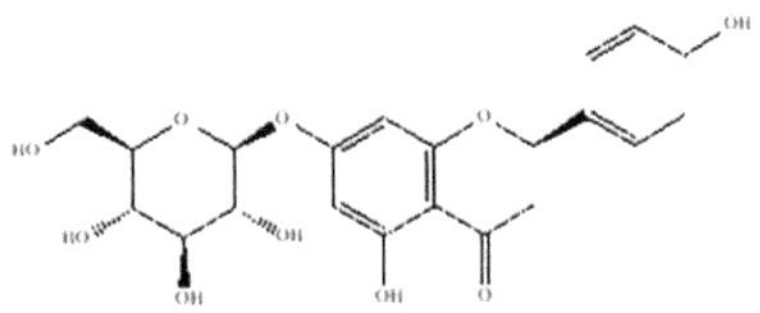
	Structure 23	Pinitol	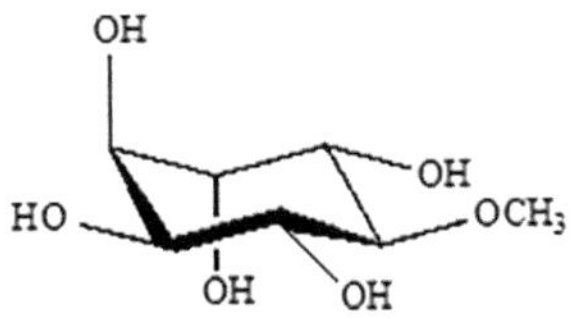

Chapter		Name	Structure
2 (*Continued*)	**Structure 24**	Sucrose	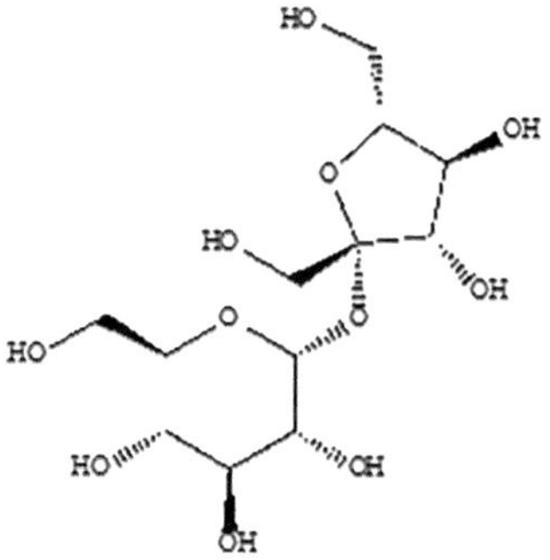
	Structure 25	Proanthocyanidin	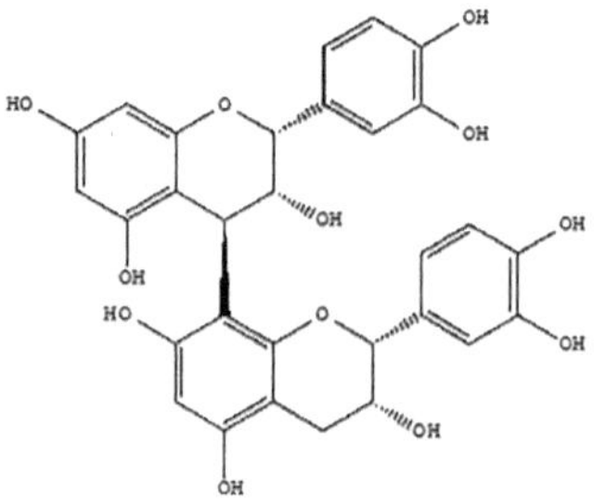
3	**Structure 26**	Acthaside (7-hydroxy-2-methyl-6-[β-galactopyranosyl-propyl]-4H-chromen-4-one) (Adapted from Amoussa, A.M., M. Bourjot, L. Lagnika, C. Vonthron-Sénécheau, and A. Sanni. 2016. Acthaside: A new chromone derivative from *Acacia ataxacantha* and its biological activities. *BMC Complement Altern Med* 16(1): 506. https://doi.org/10.1186/s12906-016-1489-y.)	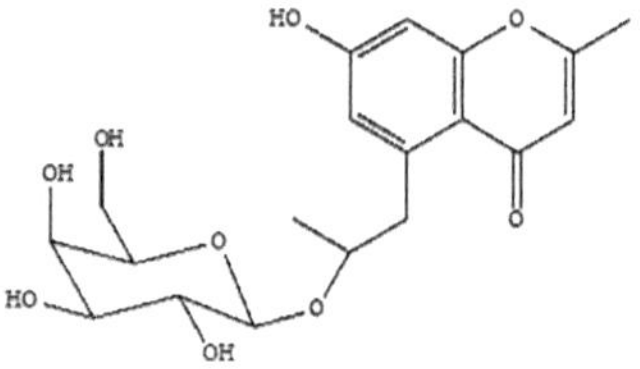
	Structure 27	Kaempferol	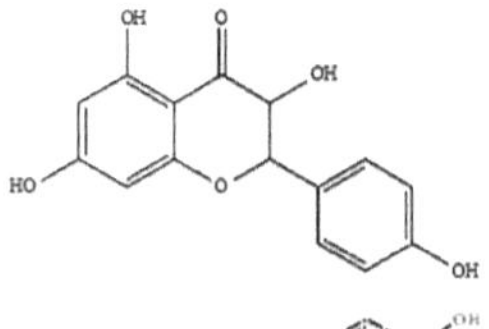
	Structure 28	Apigenin	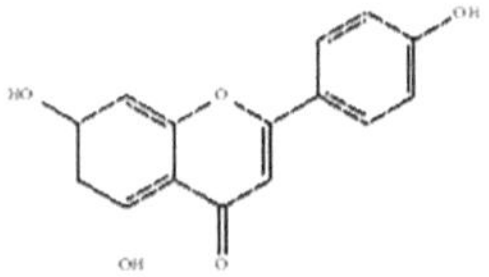
	Structure 29	3,4′,7,8-Tetrahydroxyflavanone	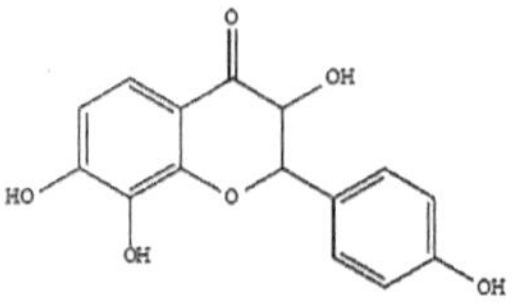

Chapter		Name	Structure
3 *(Continued)*	**Structure 30**	Teracacidin	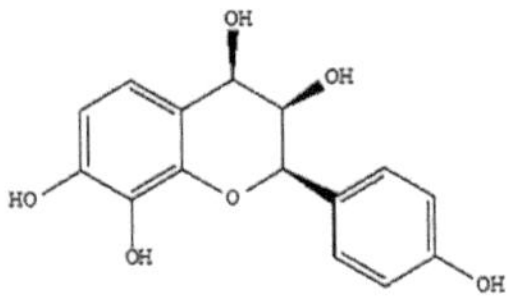
4	**Structure 31**	Hesperidin	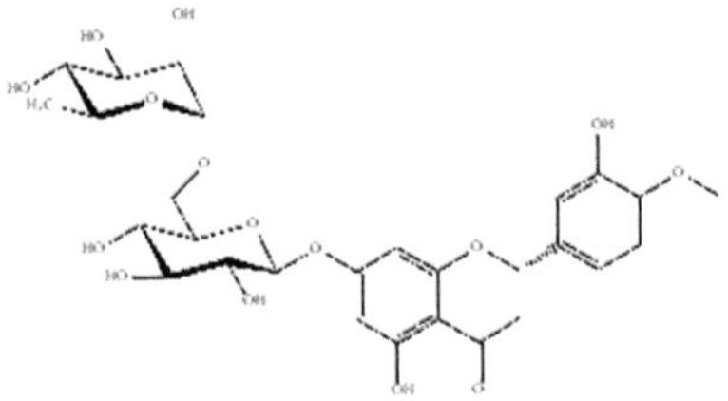
	Structure 32	7-O-Galloyl catechin	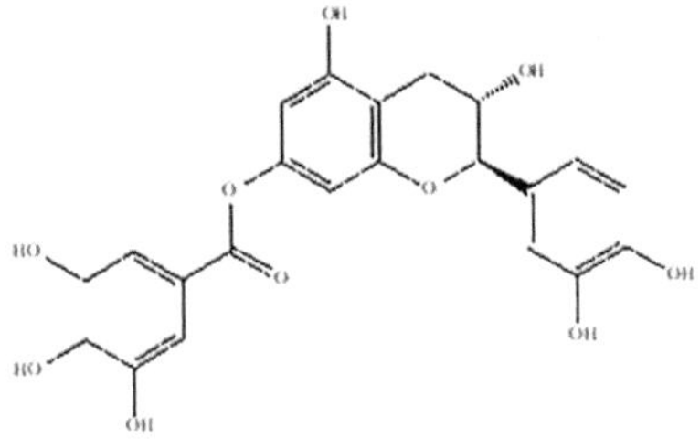
	Structure 33	cis-Verbenol	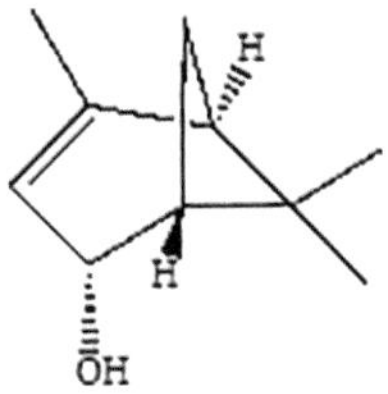
	Structure 34	Octadecyl alcohol	
	Structure 35	Phytol	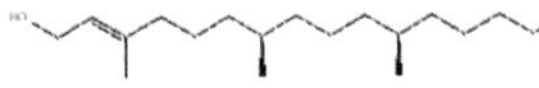
	Structure 36	Baicalin	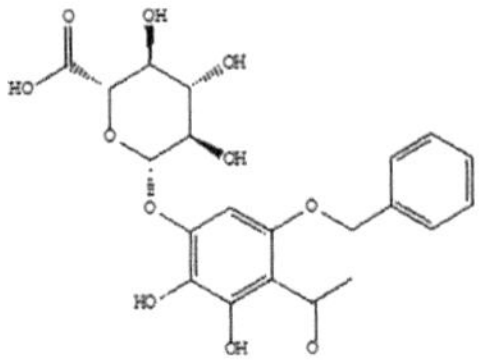
	Structure 37	Berberine	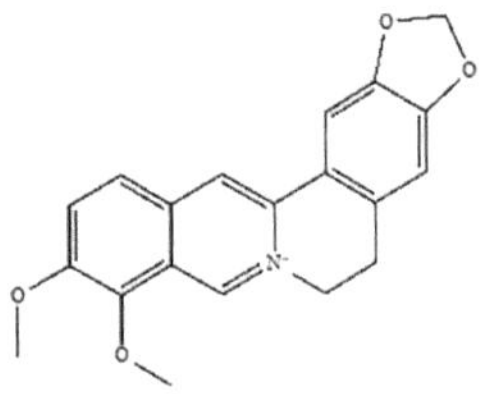
	Structure 38	Benzo(a)pyrene	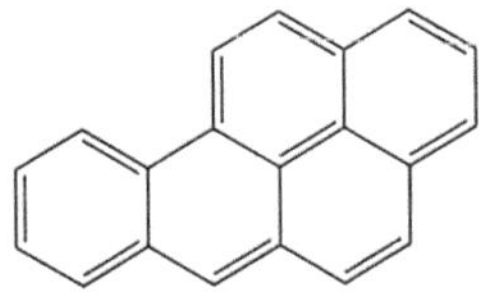

Chapter		Name	Structure
4 (*Continued*)	**Structure 39**	Naloxone	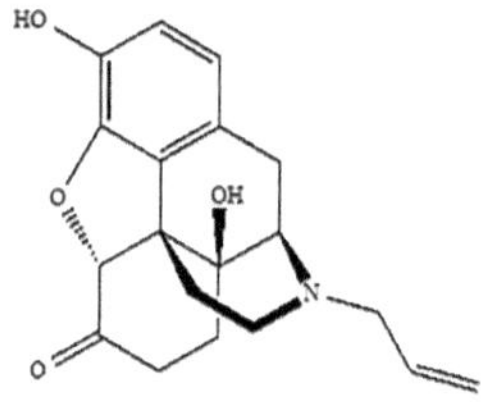
	Structure 40	3β-Acetoxy-17β-hydroxy-androst-5-ene (Adapted with permission from Chaubal, R., A.M. Mujumdar, V.G. Puranik, V.H. Deshpande, and N.R. Deshpande. 2003. Isolation and X-ray study of an anti-inflammatory active androstene steroid from *Acacia nilotica*. *Planta Med* 69(3): 287–8. https://doi.org/10.1016/S0040-4020(01)80017-X.)	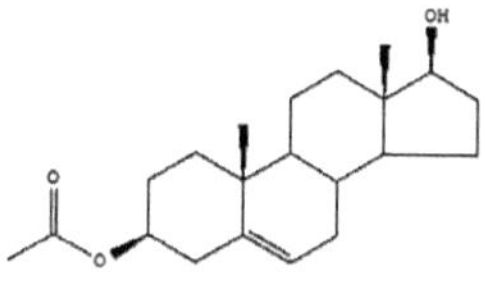
	Structure 41	Hydroxytyrosol acetate	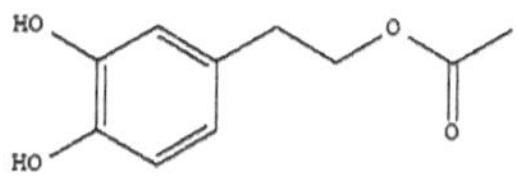
	Structure 42	Quinic acid	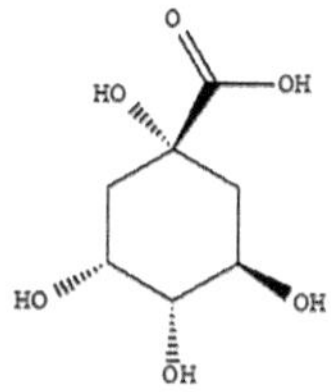
	Structure 43	Caffeoylmalic acid	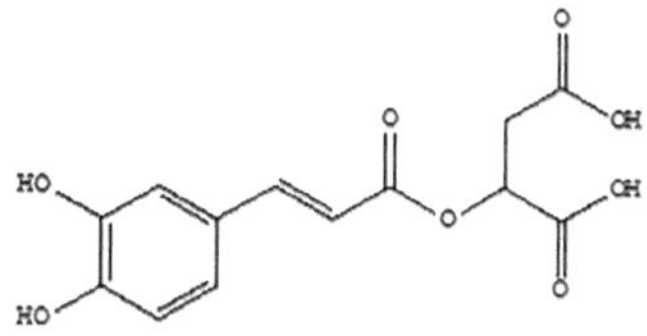
	Structure 44	Heptyl valerate	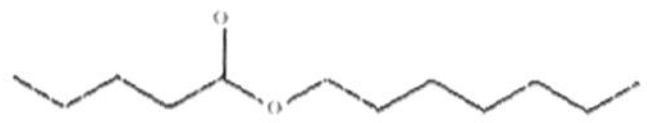
	Structure 45	Nonadecane	
	Structure 46	Melanoxetin (3,7,8,3′,4′-pentahydroxyflavone)	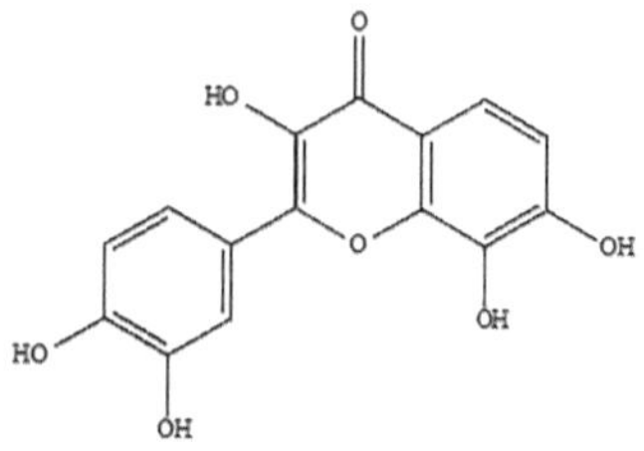

Chapter	Name		Structure
4 (*Continued*)	**Structure 47**	Lupeol	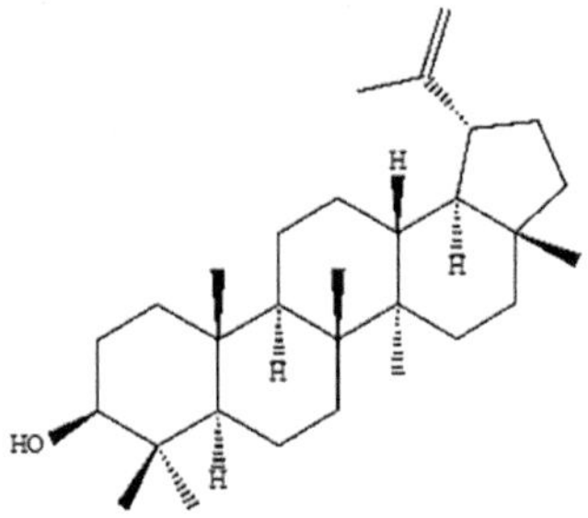
	Structure 48	α-Amyrin	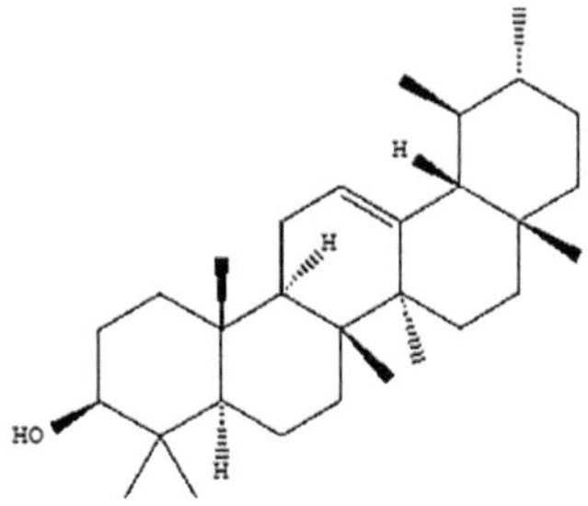
	Structure 49	β-Amyrin	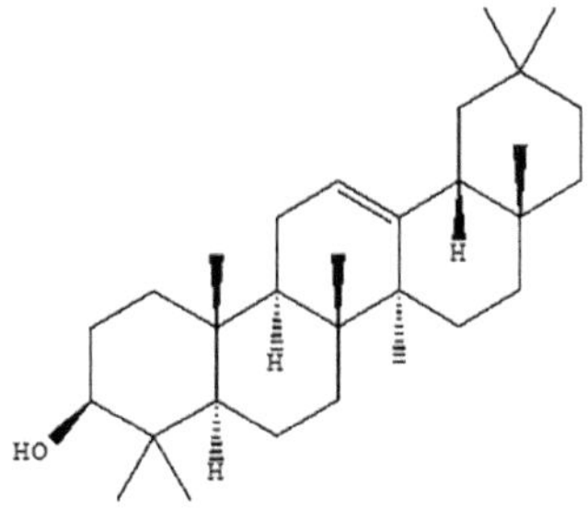
5	**Structure 50**	Rutin	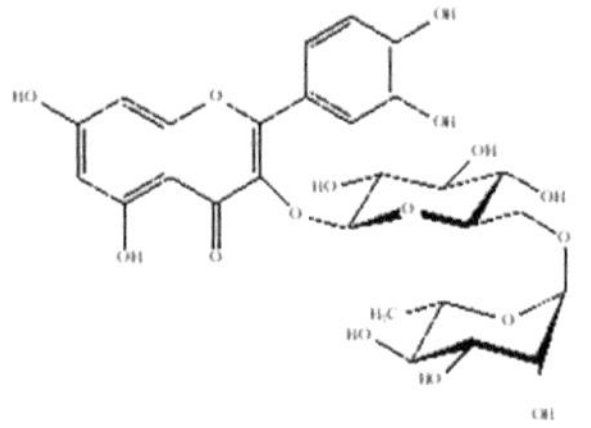
	Structure 51	Quercetin	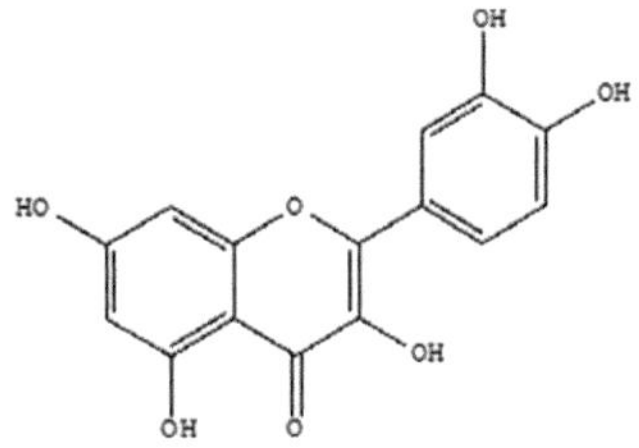

Chapter		Name	Structure
5 *(Continued)*	**Structure 52**	N,N-Dimethyltryptamine-N-oxide	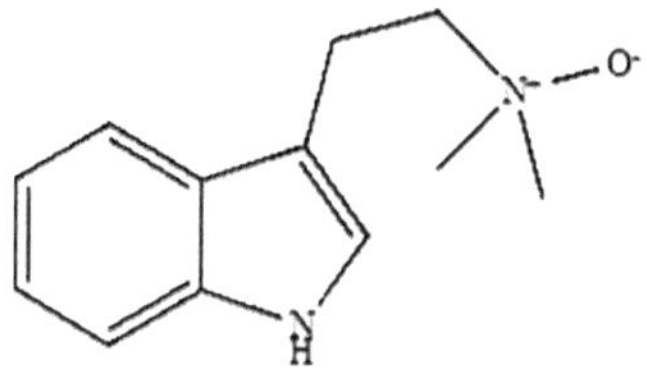
	Structure 53	N-Chloromethyl-N, N-dimethyltryptamine	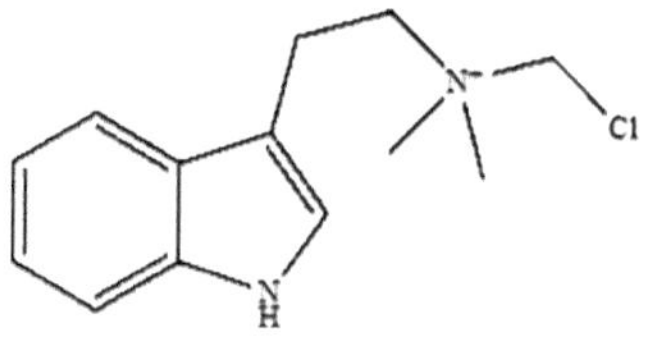
	Structure 54	Protocatechuic acid	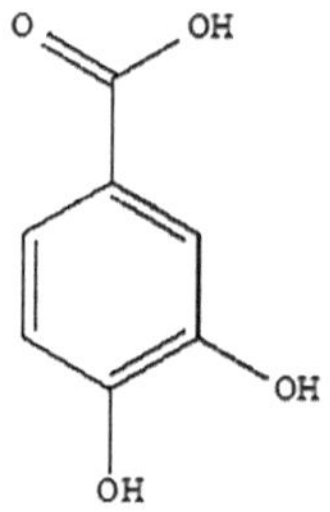
	Structure 55	Caffeic acid	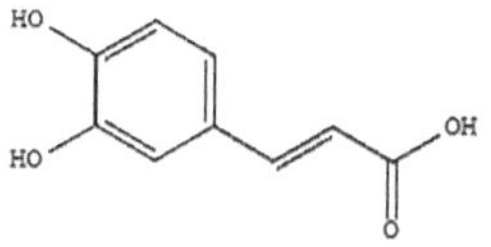
	Structure 56	Ellagic acid	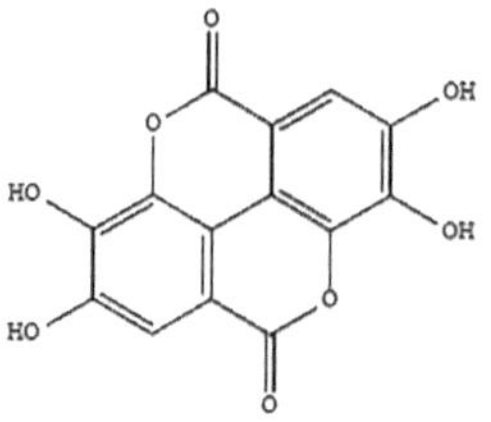
	Structure 57	Curcumin	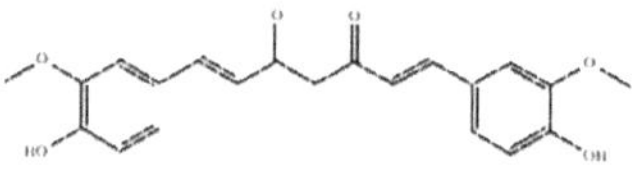
	Structure 58	Catechin-3-O-gallate	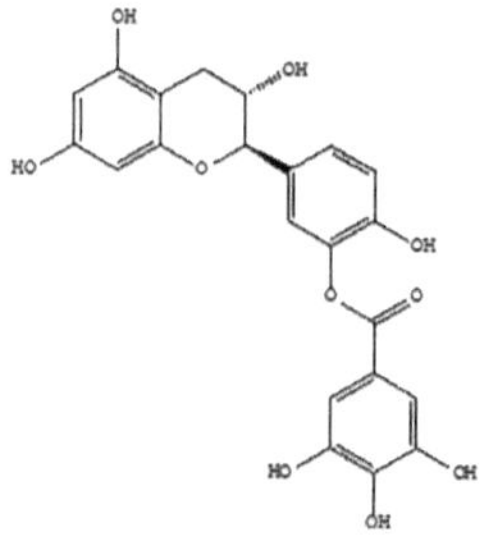
	Structure 59	Myricetin	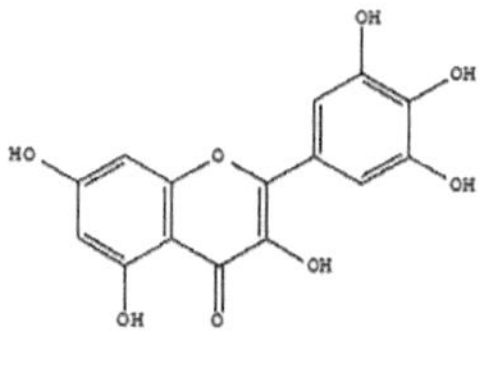

Chapter		Name	Structure
5 *(Continued)*	**Structure 60**	5-Fluorouracil	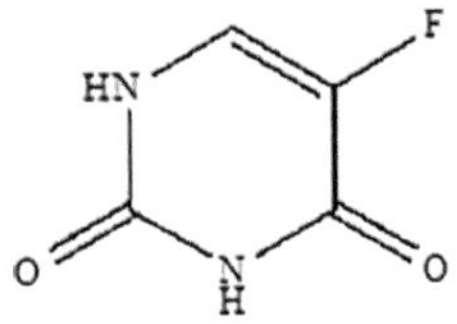
	Structure 61	Polyacanthoside A (Adapted with permission from Fotso, W.G., J. Na-Iya, T.A. Mbaveng, et al. 2019. Polyacanthoside A, a new oleanane-type triterpenoid saponin with cytotoxic effects, is found in the leaves of *Acacia polyacantha* (Fabaceae). *Nat Prod Res* 33(24): 3521–6. https://doi.org/10.1080/14786419.2018.1486312.)	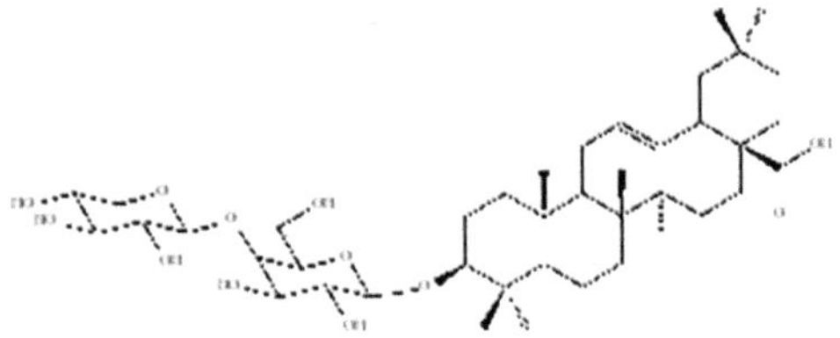
	Structure 62	Erythrodiol	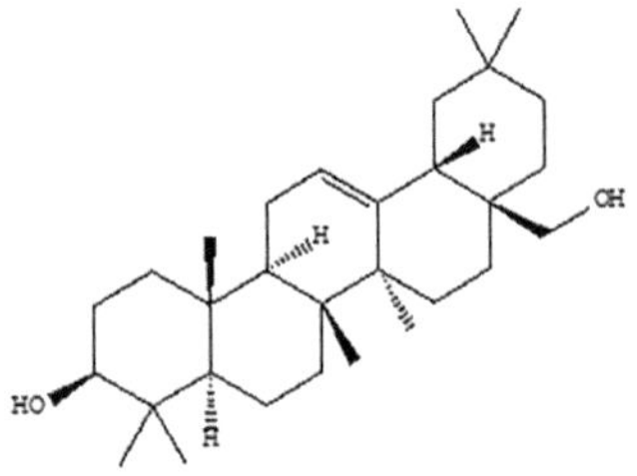
	Structure 63	Quercetin-3-O-α-L-rhamnoside	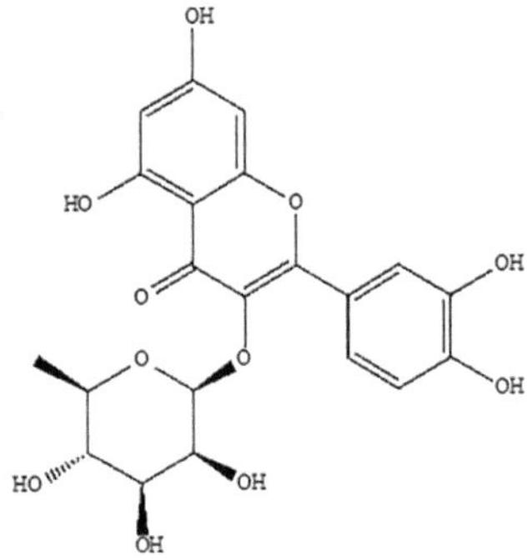
	Structure 64	Myricetin-3-O-α-L-rhamnoside	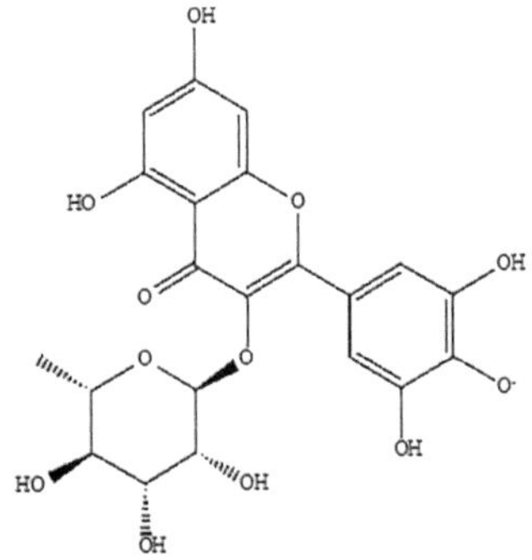

Chapter		Name	Structure
5 *(Continued)*	**Structure 65**	Melacacidin	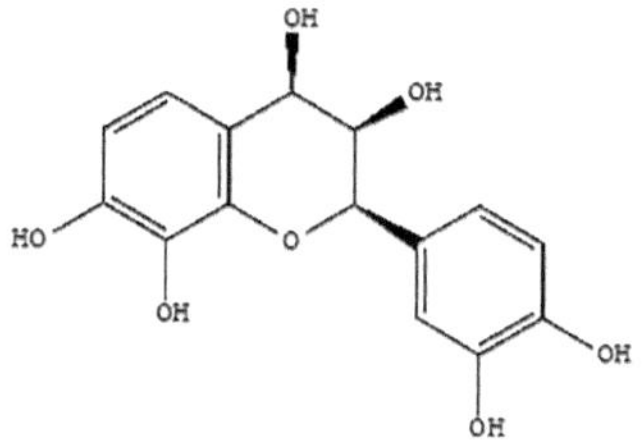
	Structure 66	7,8,3′,4′-Tetrahydroxyflavanone	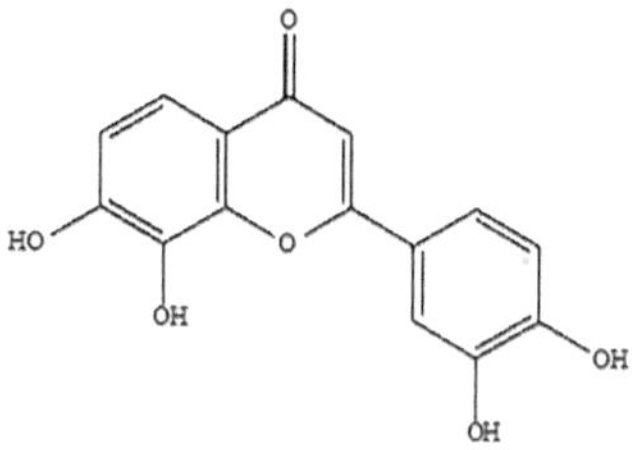
	Structure 67	(20R)/(20S)-3-Oxolupan-30-al	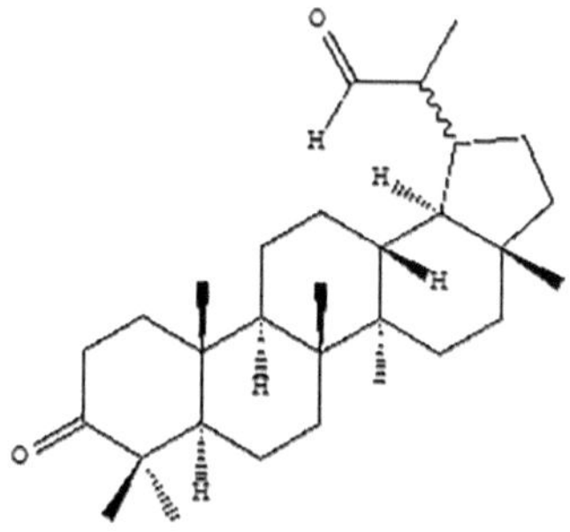
	Structure 68	(20R)-28-Hydroxylupen-30-al-3-one	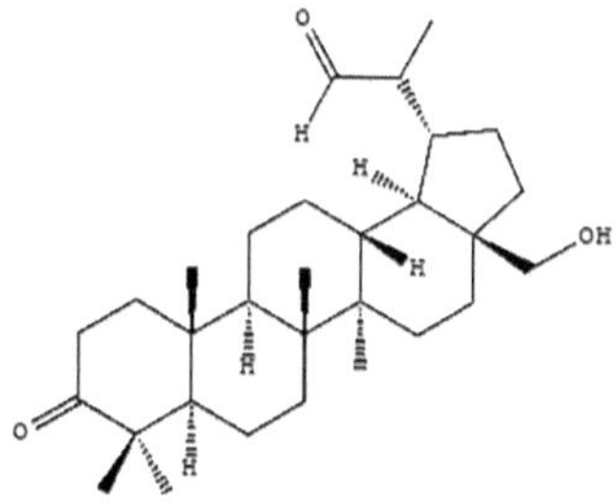
	Structure 69	(20S)-3β-Hydroxylupan-30-al	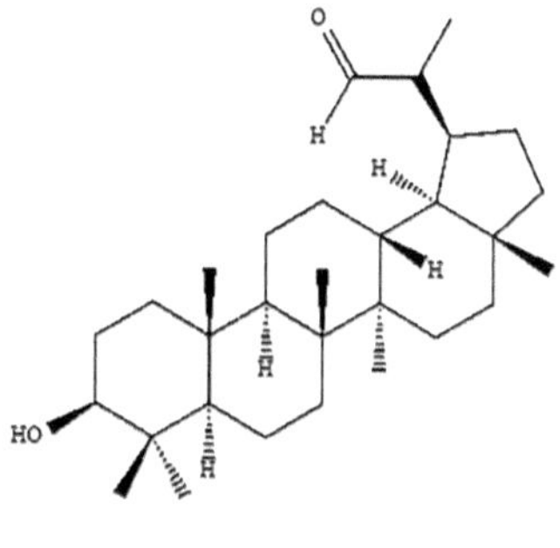

Chapter	Name	Structure
5 (*Continued*)	**Structure 70** 30-Hydroxylup-20-(29)-en-3-one	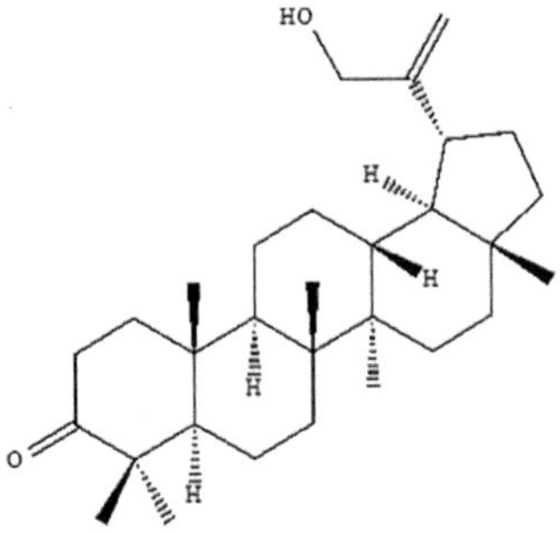
	Structure 71 30-Hydroxylup-20-(29)-en-3β-ol	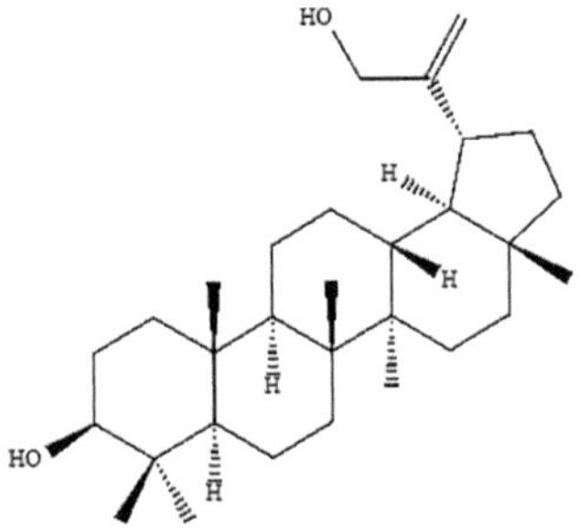
	Structure 72 Atranorin	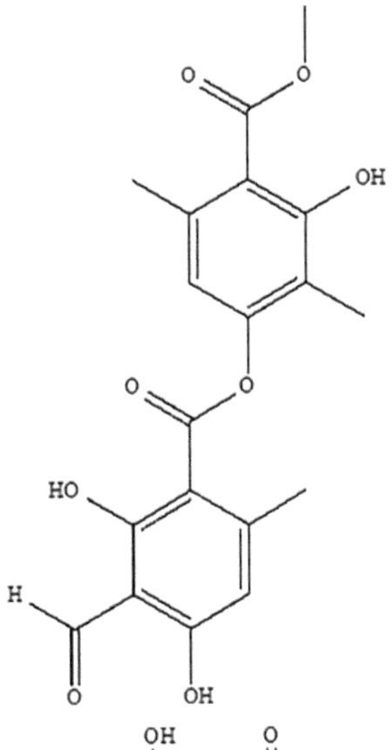
	Structure 73 Methyl 2,4-dihydroxy-3,6 dimethyl benzoate	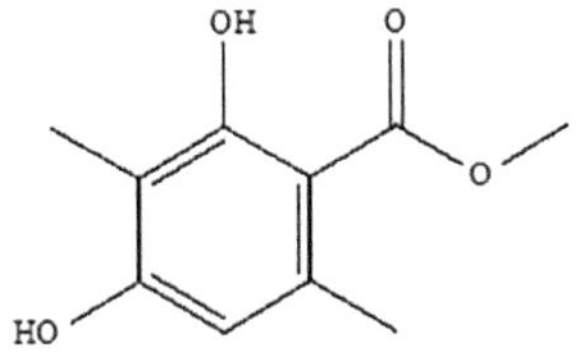
	Structure 74 Linoleic acid	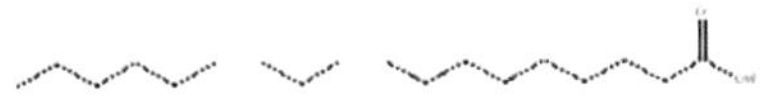

Chapter		Name	Structure
5 *(Continued)*	**Structure 75**	Cyclopside 1	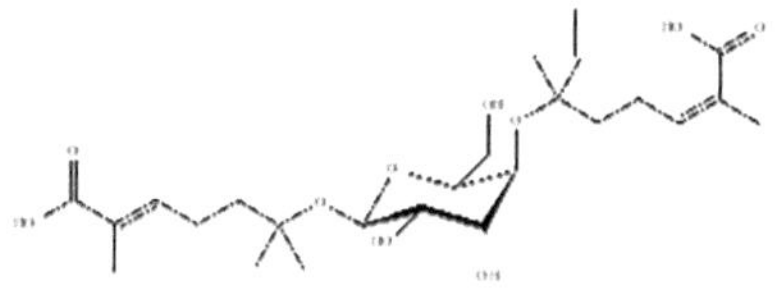
	Structure 76	Cyclopside 2	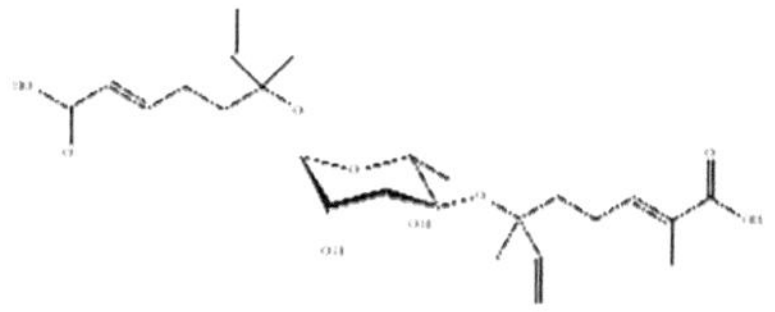
	Structure 77	seco-Oxacassane 1	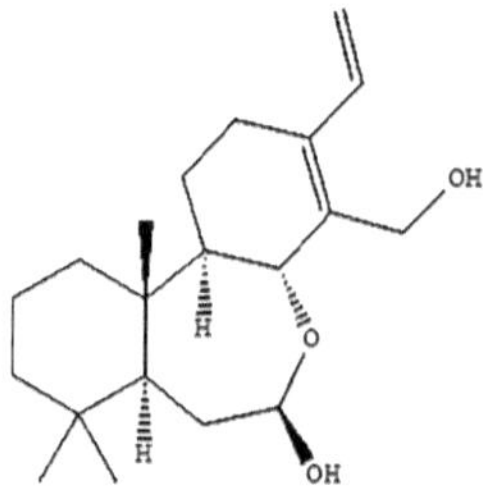
	Structure 78	seco-Oxacassane 2	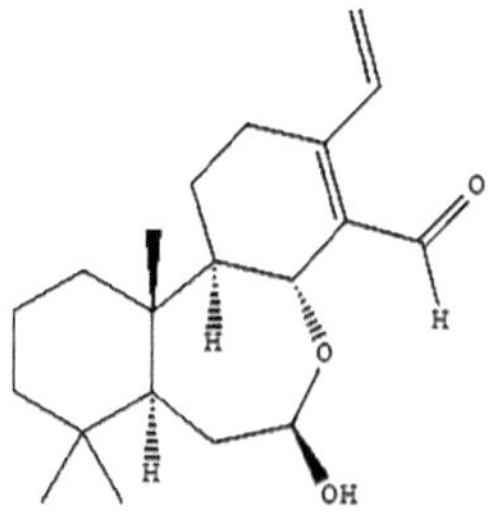
	Structure 79	seco-Oxacassane 3	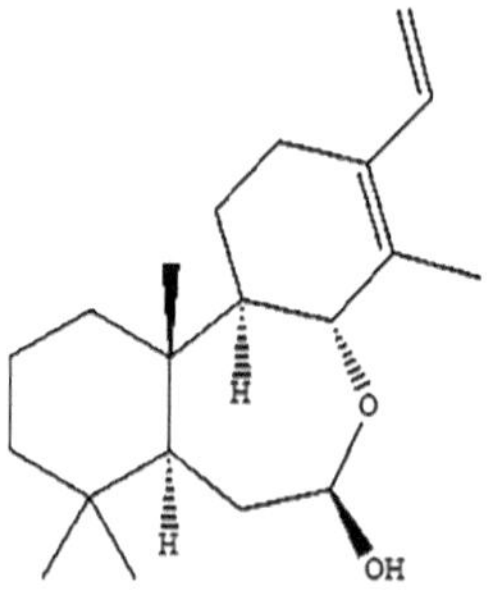
	Structure 80	Ligulataside a (Adapted with permission from Jæger, D., C.P. Ndi, C. Crocoll, et al. 2017. Isolation and structural characterization of echinocystic acid triterpenoid saponins from the Australian medicinal and food plant *Acacia ligulata*. *J Nat Prod* 80(10): 2692–8. Copyright 2017 American Chemical Society. https://doi.org/10.1021/acs.jnatprod.7b00437.)	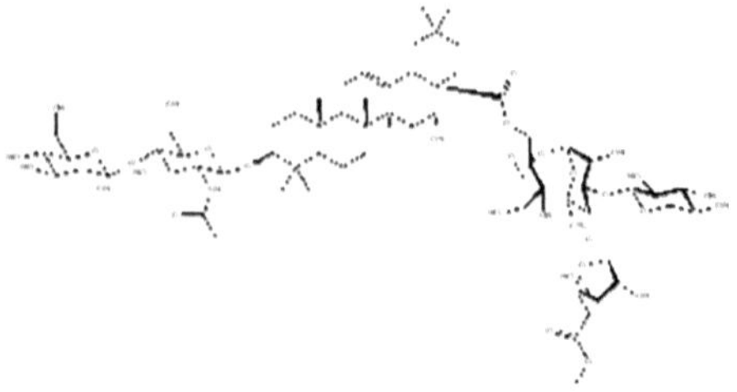

Chapter		Name	Structure
5 *(Continued)*	**Structure 81**	Ligulataside b (Adapted with permission from Jæger, D., C.P. Ndi, C. Crocoll, et al. 2017. Isolation and structural characterization of echinocystic acid triterpenoid saponins from the Australian medicinal and food plant *Acacia ligulata*. *J Nat Prod* 80(10): 2692–8. Copyright 2017 American Chemical Society. https://doi.org/10.1021/acs.jnatprod.7b00437.)	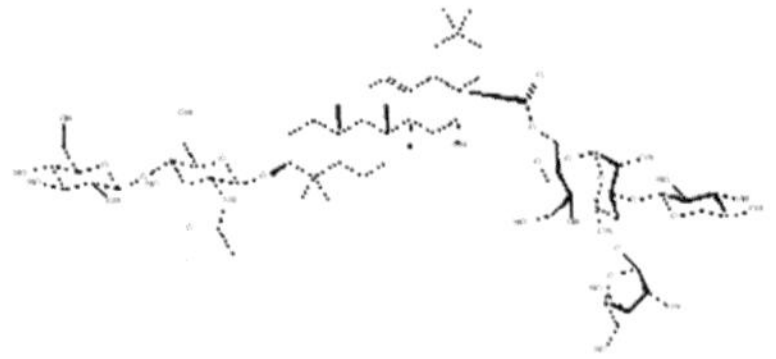
	Structure 82	Avicin D (Adapted from Haridas, V., Z.X. Xu, D. Kitchen, A. Jiang, P. Michels, and J.U. Gutterman. 2011. The anticancer plant triterpenoid, avicin D, regulates glucocorticoid receptor signaling: Implications for cellular metabolism. *PLoS One* 6(11): e28037. CC BY; https://doi.org/10.1371/journal.pone.0028037.)	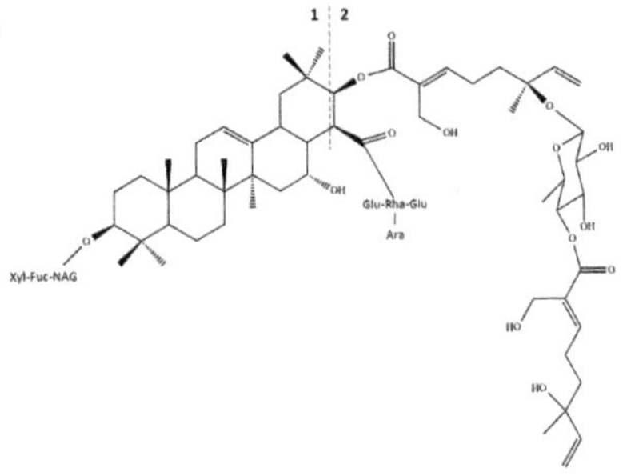
6	**Structure 83**	1,2-Benzenedicarboxylic acid mono (2-ethylhexyl) ester	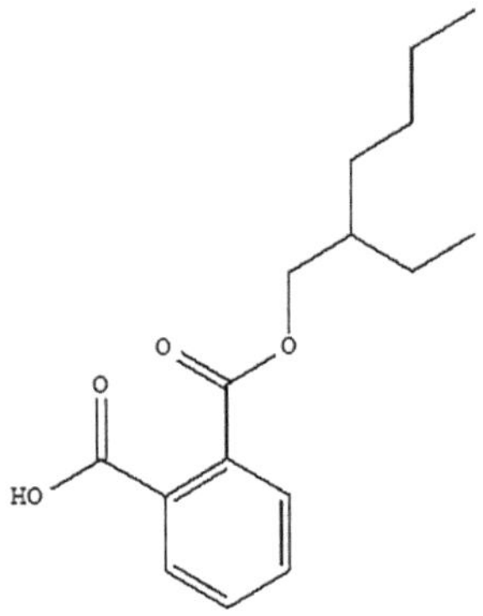
	Structure 84	2,6-Dimethyl-N-(2-methyl-α-phenylbenzyl) aniline	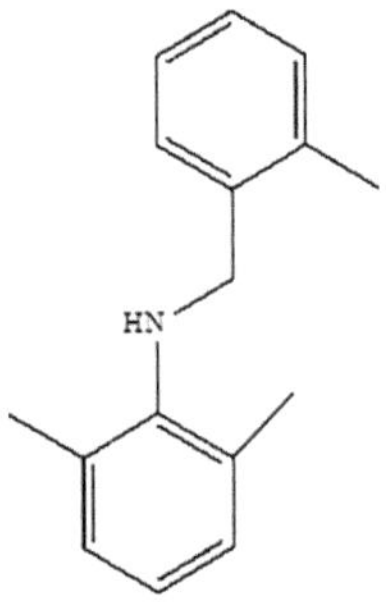
	Structure 85	Squalene	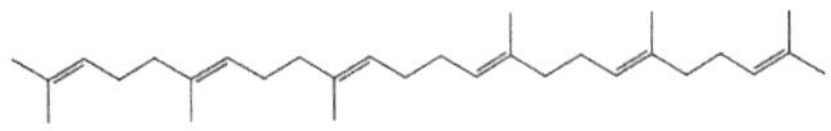

Chapter		Name	Structure
6 (*Continued*)	**Structure 86**	N-Methyl β-phenylethylamine	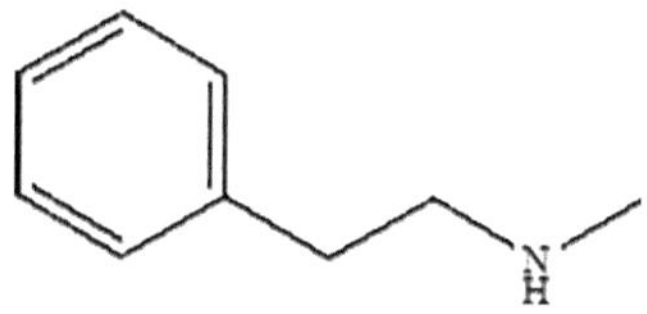
	Structure 87	N-Methyl tyramine	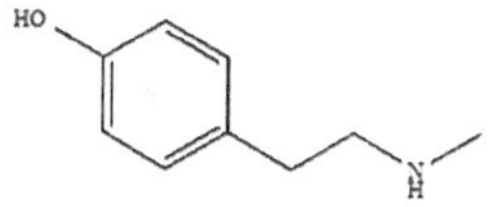
	Structure 88	Psilocin	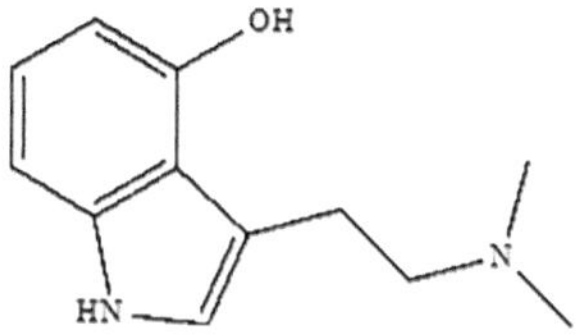
	Structure 89	Psilocybin	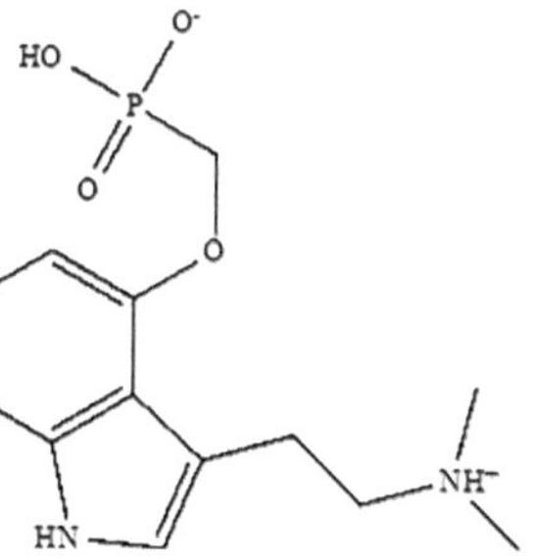
	Structure 90	Harmine	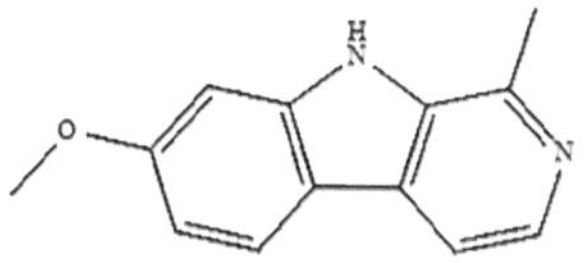
	Structure 91	Harmaline	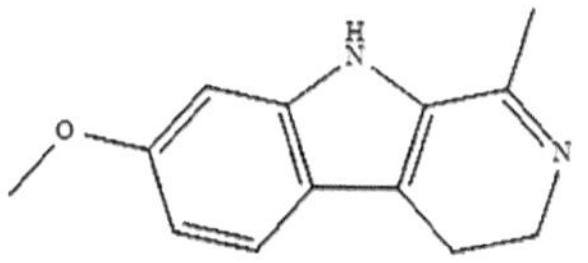
7	**Structure 92**	Robinetinidol	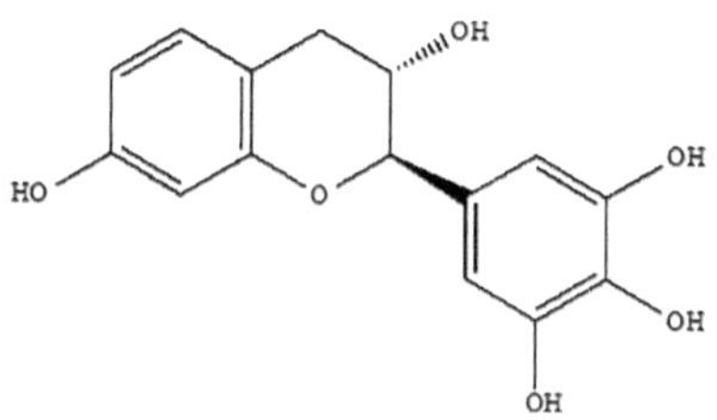
	Structure 93	Fisetinidol	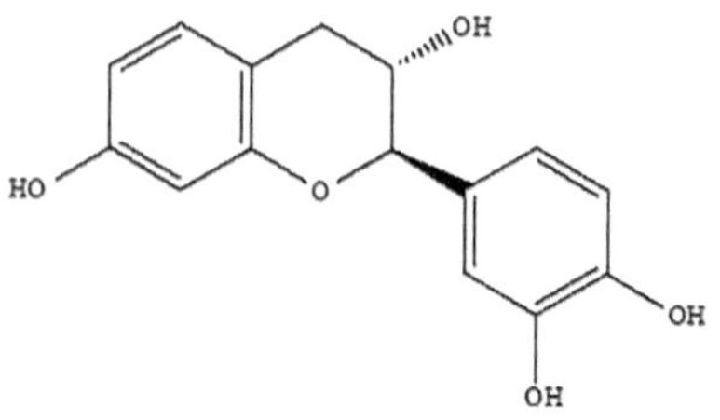

Chapter	Name		Structure
8	**Structure 94**	Resveratrol	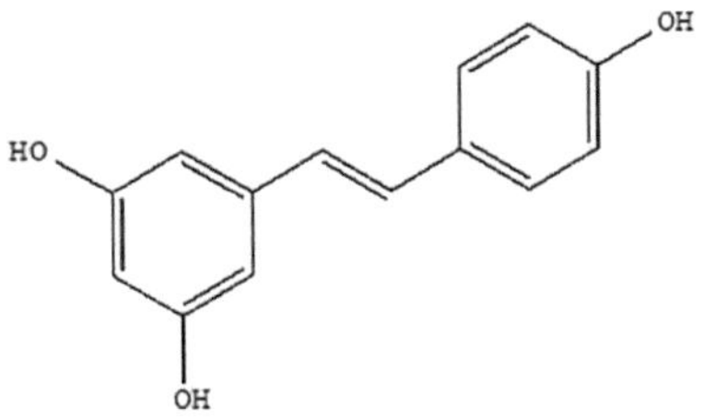
	Structure 95	Tetrahydrocannabivarin	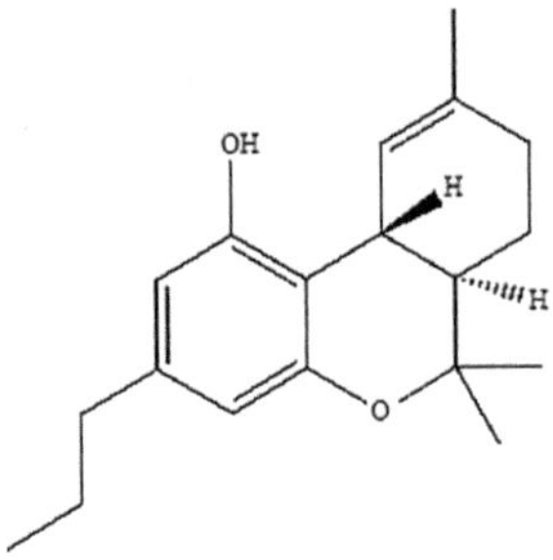
	Structure 96	Tetrahydrocannabinol	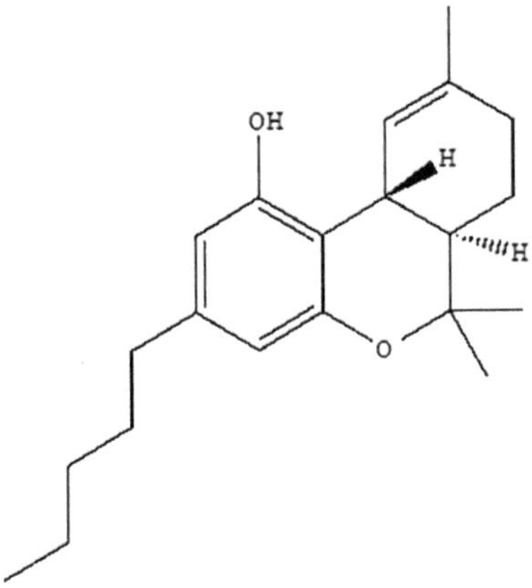
	Structure 97	Concinnoside A	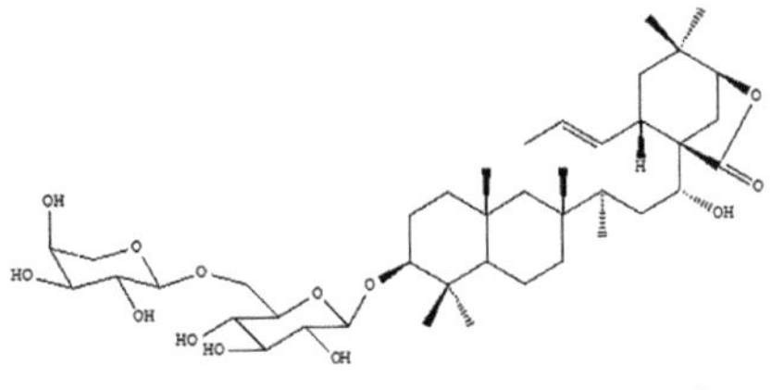
	Structure 98	Concinnoside B	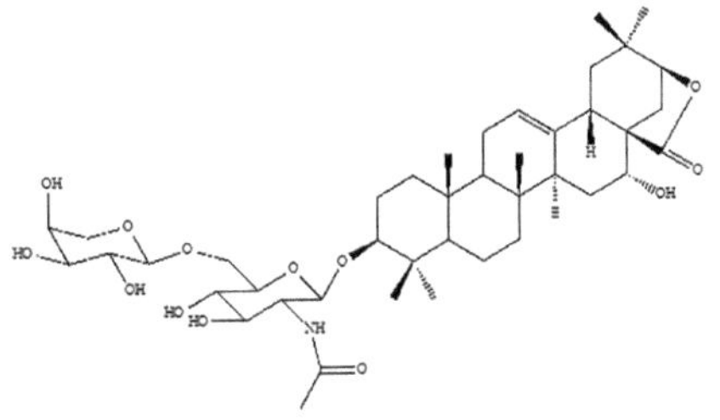

Chapter		Name	Structure
8 *(Continued)*	**Structure 99**	Concinnoside C	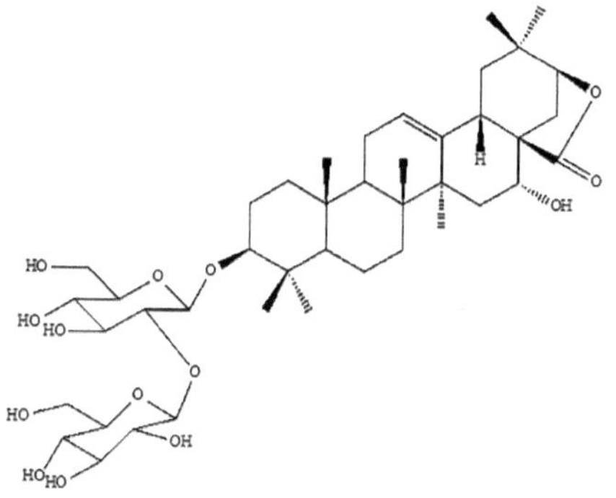
	Structure 100	Concinnoside D	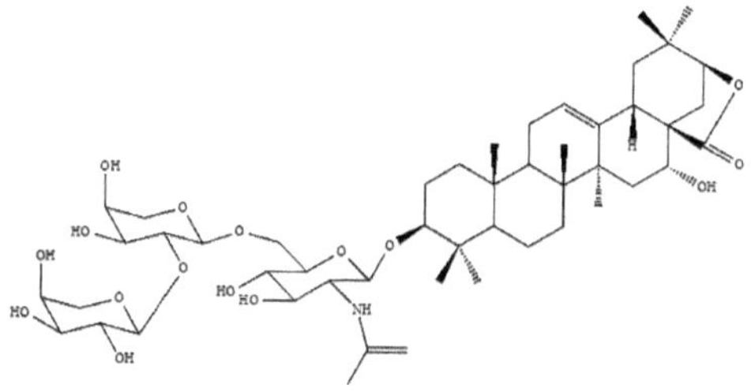
	Structure 101	Concinnoside E	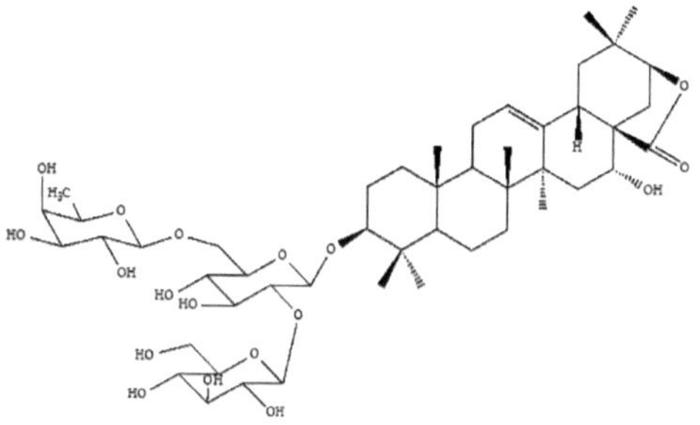
	Structure 102	Acaciaside	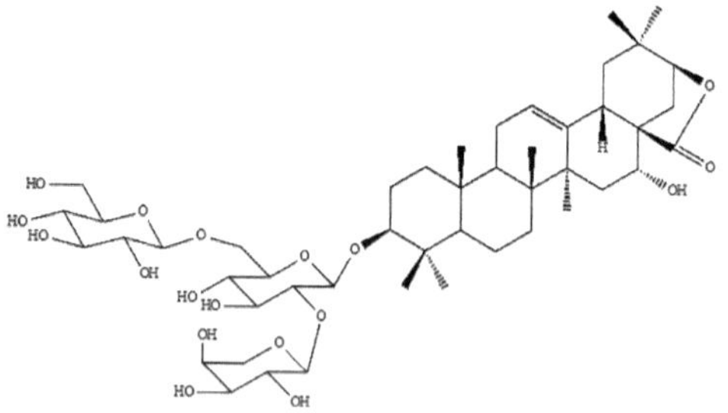
	Structure 103	Julibroside A1	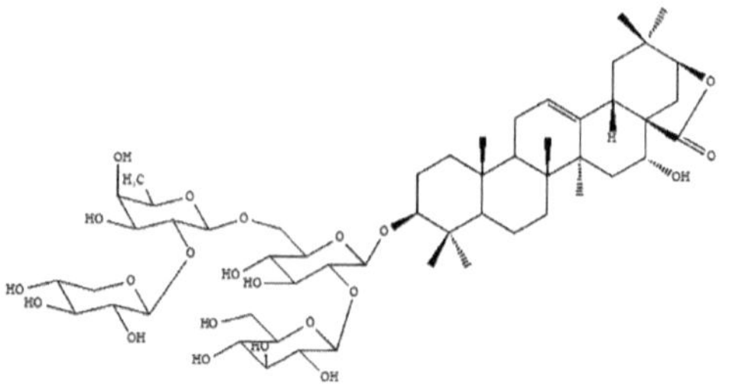

Chapter	Name		Structure
8 *(Continued)*	**Structure 104**	Julibroside A3	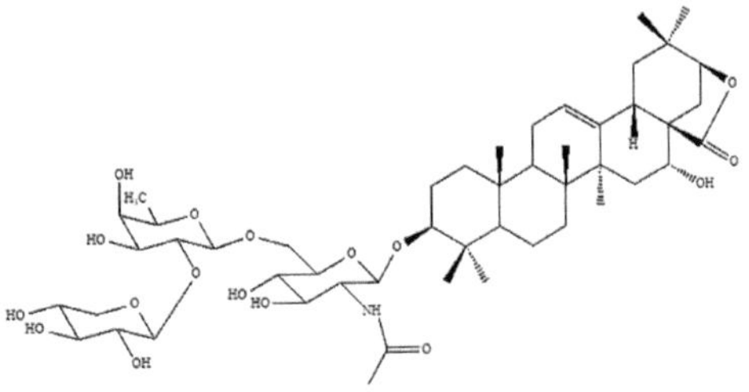
	Structure 105	Albiziasaponin C	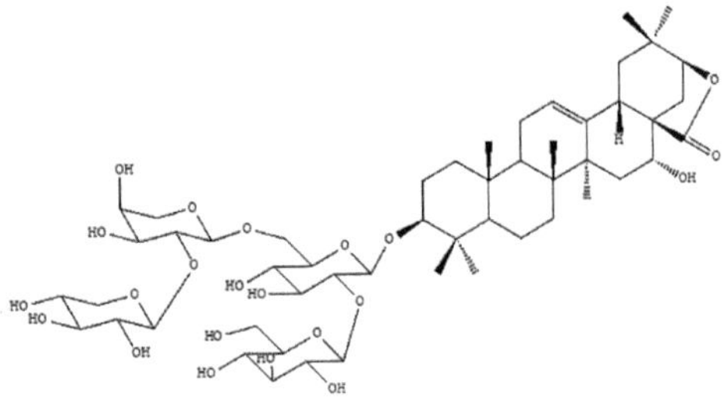
	Structure 106	Acacic acid lactone	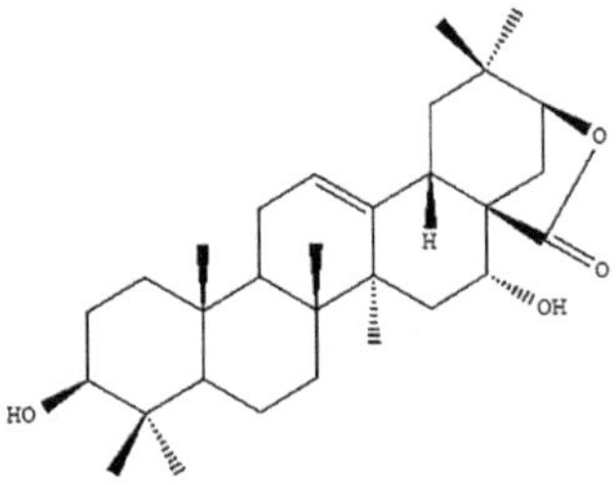
	Structure 107	Gallocatechin	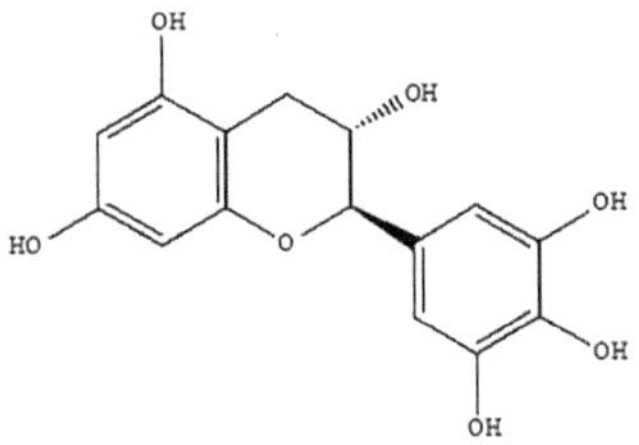
	Structure 108	Fisetinidol-(4α-8)-catechin	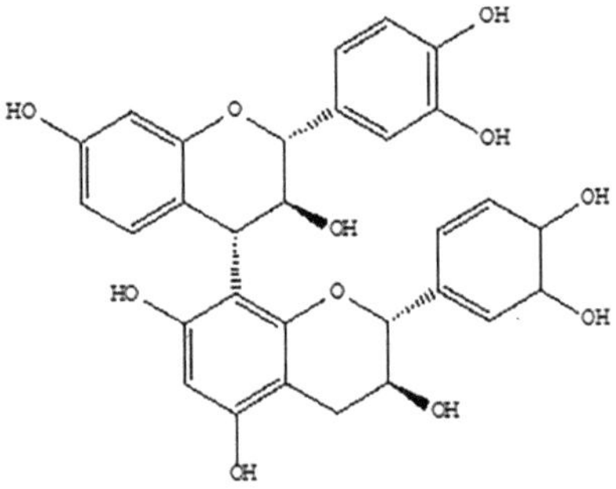

Chapter		Name	Structure
8 *(Continued)*	**Structure 109**	Robinetinidol-(4α-8)-catechin	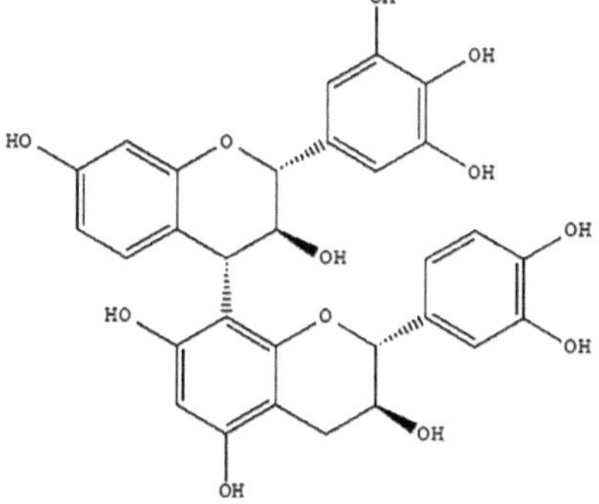
	Structure 110	Robinetinidol-(4α-8)-gallocatechin	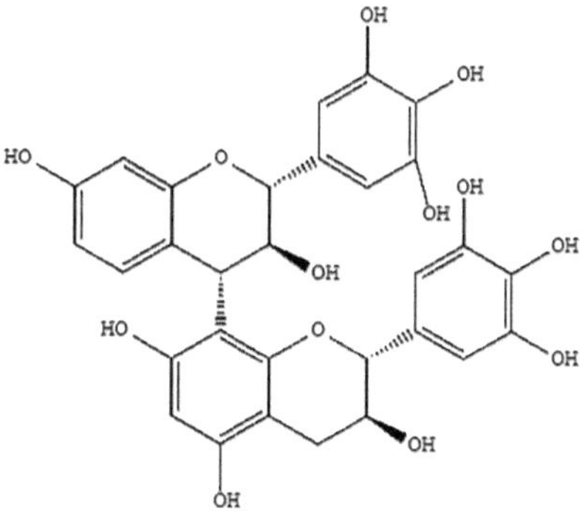
	Structure 111	Leucorobinetinidin	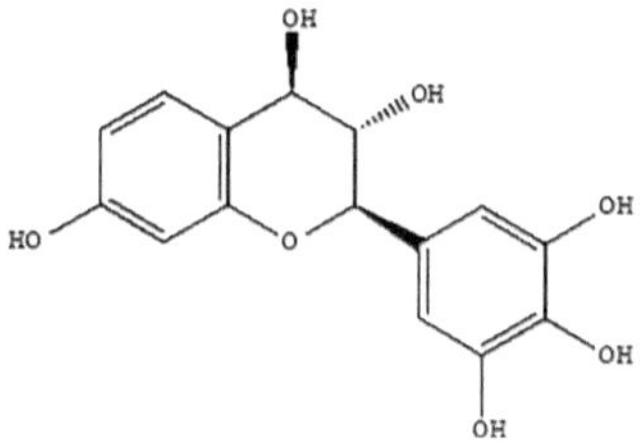
	Structure 112	Myricetin	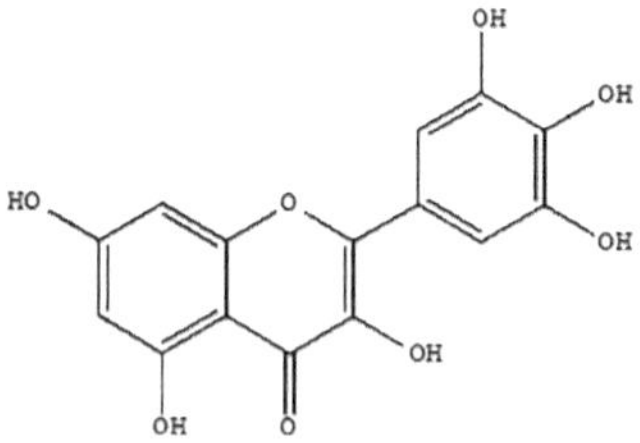
	Structure 113	Butin	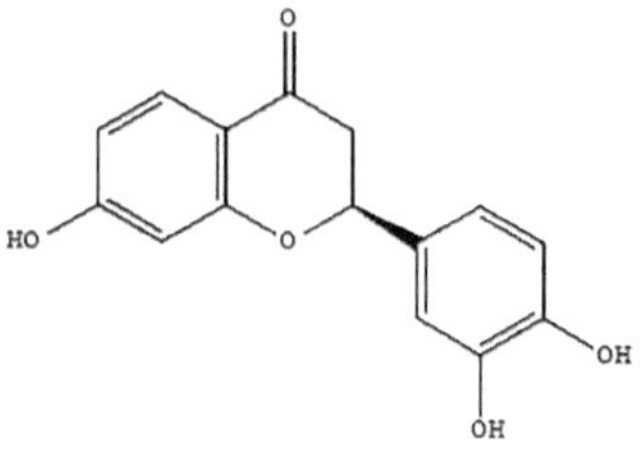

Chapter	Name		Structure
8 *(Continued)*	**Structure 114**	Butein	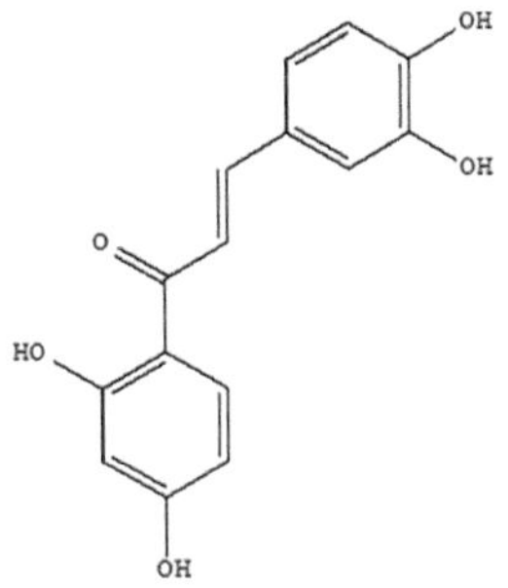
	Structure 115	Robtein	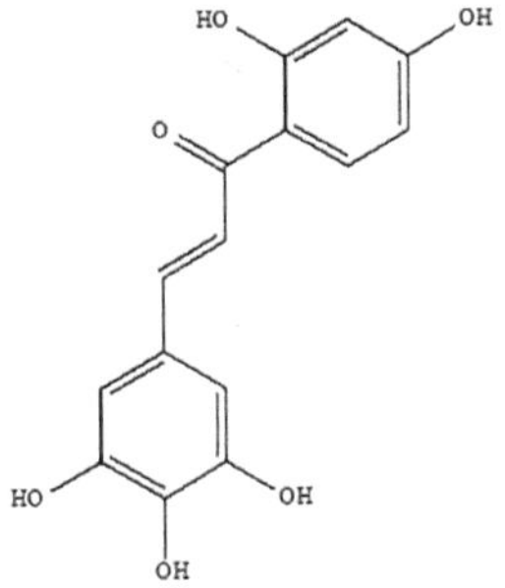
	Structure 116	Fustin	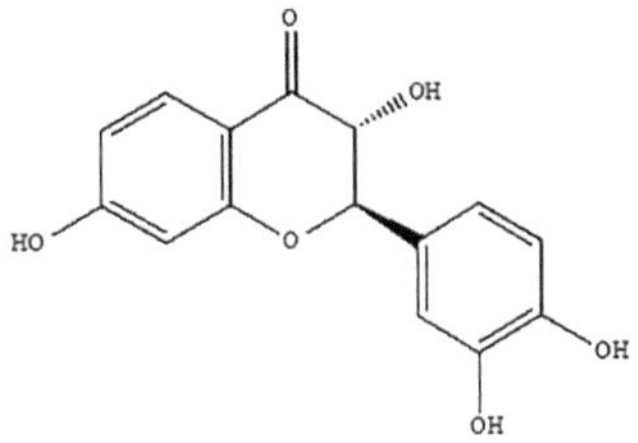
	Structure 117	Dihydrorobinetin	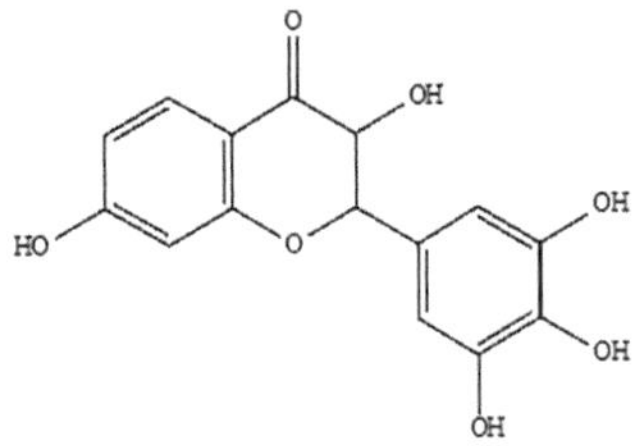
	Structure 118	Fisetin	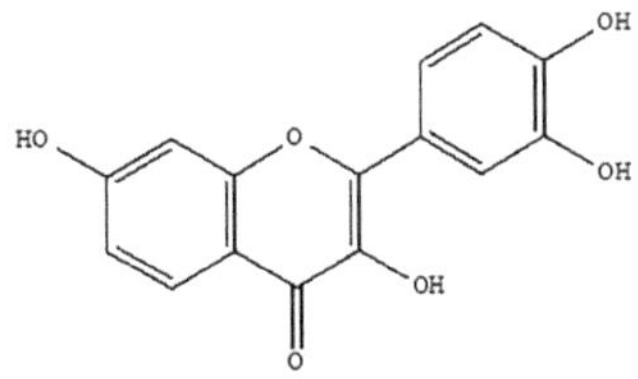

Chapter		Name	Structure
8 *(Continued)*	**Structure 119**	Robinetin	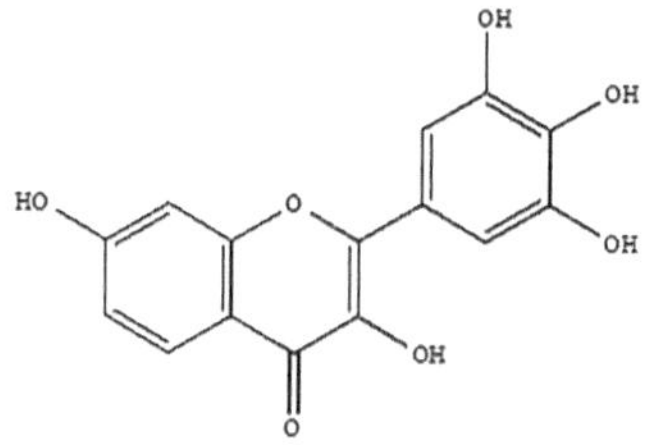
	Structure 120	Luteolin	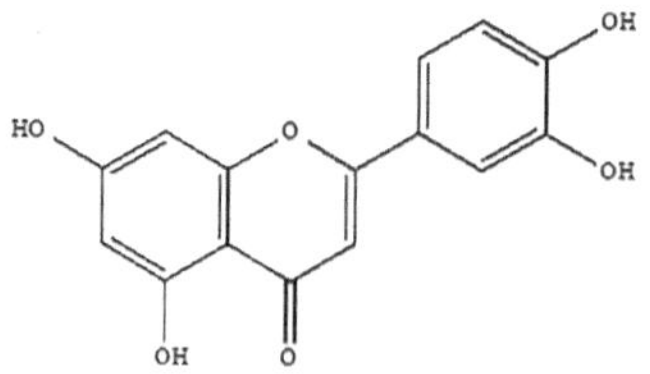
	Structure 121	Serotonin	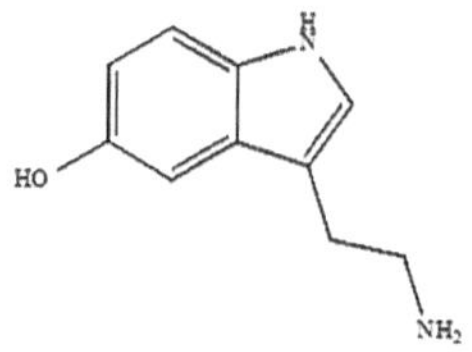
	Structure 122	(R)-2,5-Dimethoxy-4-iodoamphetamine [(R)-DOI]	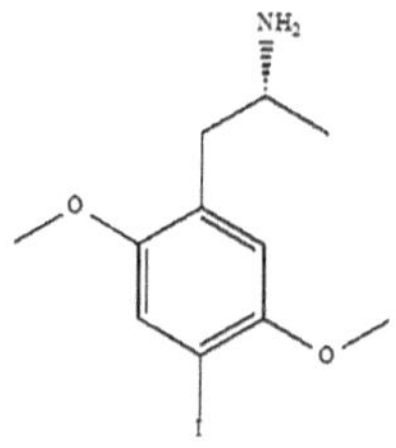
9	**Structure 123**	Streptozotocin	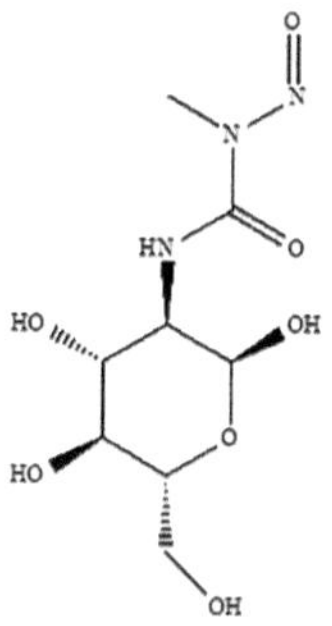
	Structure 124	Alloxan	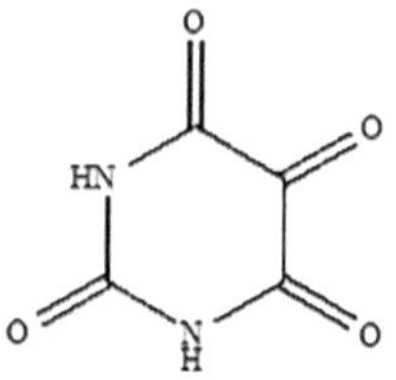
	Structure 125	Malondialdehyde	

Chapter	Name		Structure
9 *(Continued)*	**Structure 126**	d-Pinitol	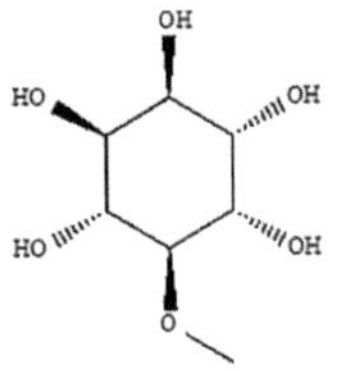
	Structure 127	Catechin-5-galloyl ester	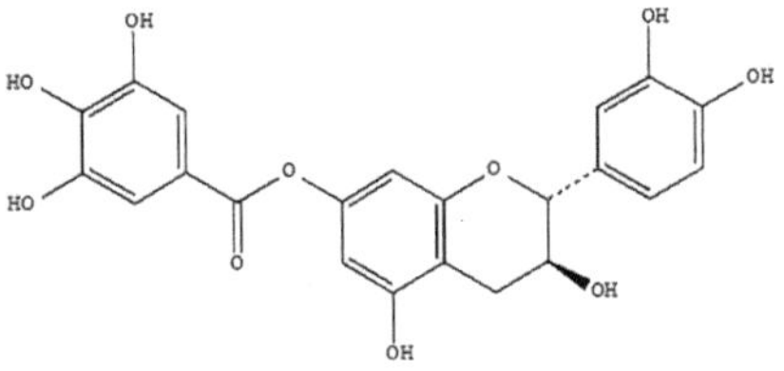
	Structure 128	m-Digallic acid	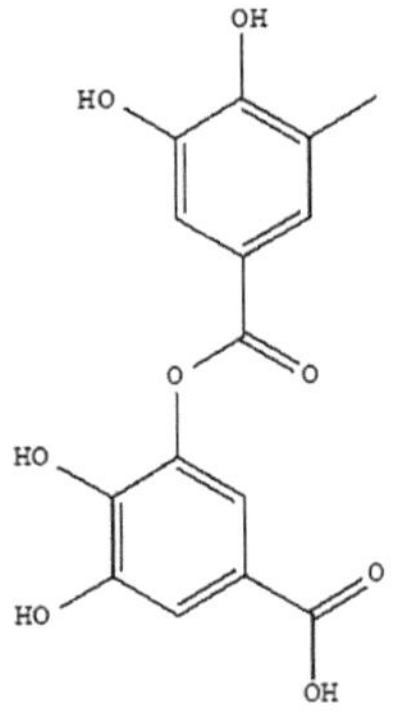
	Structure 129	Chlorogenic acid	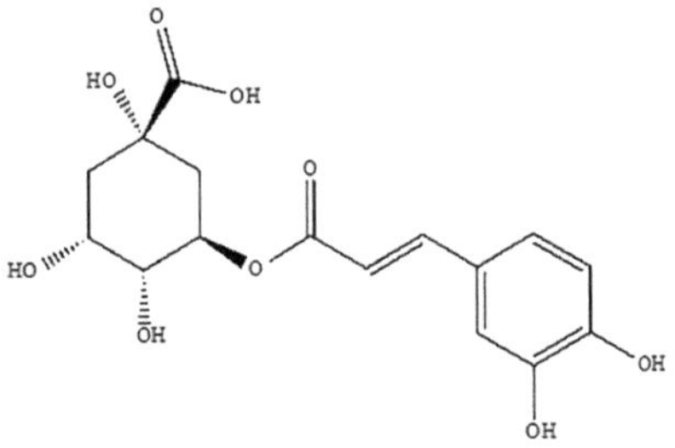
	Structure 130	Glutathione	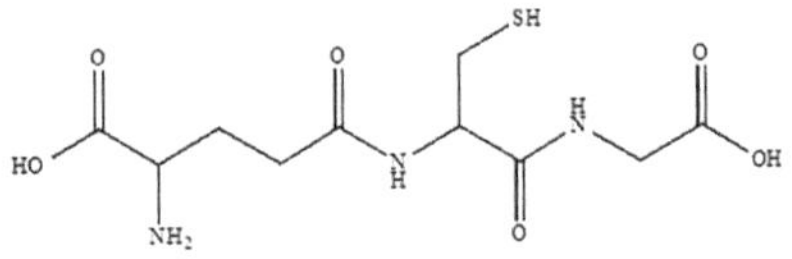
10	**Structure 131**	8-Hydroxy-2′-deoxyguanosine	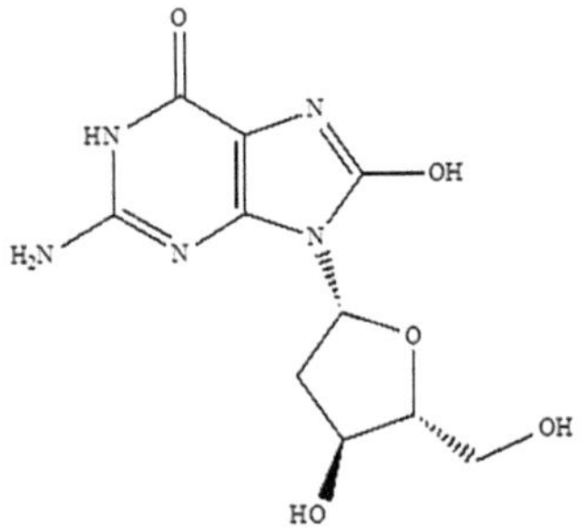

Chapter	Name	Structure
10 *(Continued)*	**Structure 132** Lidocaine	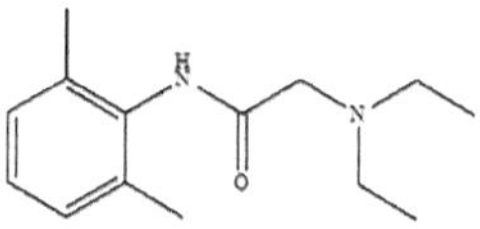
	Structure 133 Prilocaine	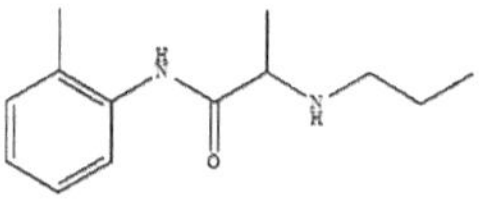
	Structure 134 Bupivacaine	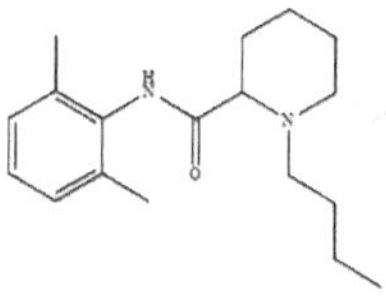
	Structure 135 2,6-Dimethoxy-1,4-benzoquinone	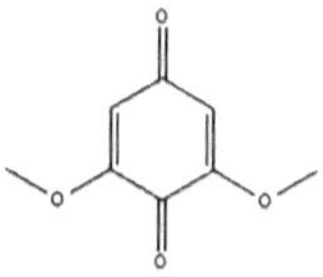
	Structure 136 Acamelin (6-methoxy-2-methyl-3, 5-dihydrobenzofurano-4,7-dion)	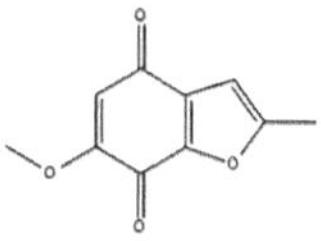
	Structure 137 7,12-Dimethylbenz(a)anthracene (DMBA)	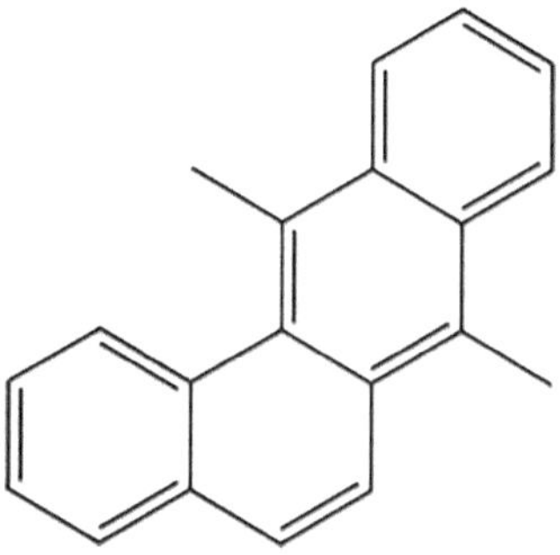
	Structure 138 Betulinic acid	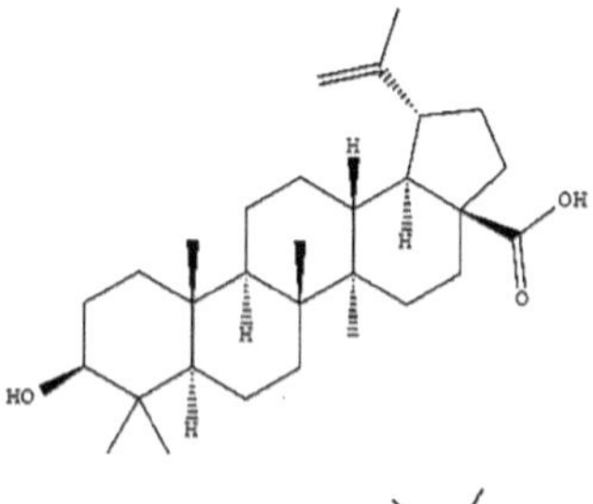
	Structure 139 Tryptamine trimethylammonium (2-(3-indolyl) ethyltrimethylammonium)	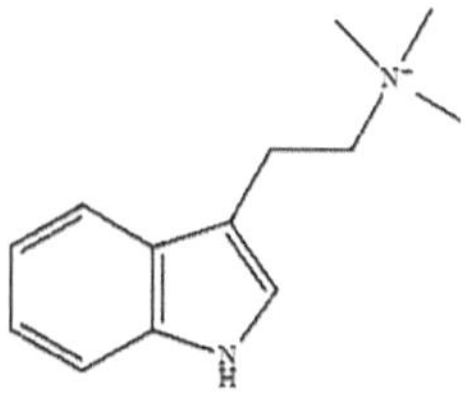

Chapter		Name	Structure
10 *(Continued)*	**Structure 140**	5-Methoxy-dimethyltryptamine	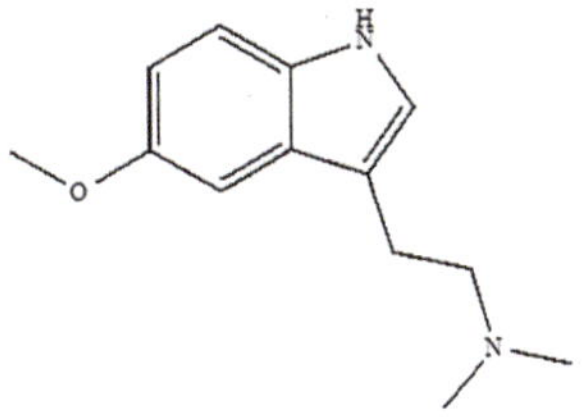
	Structure 141	Melatonin	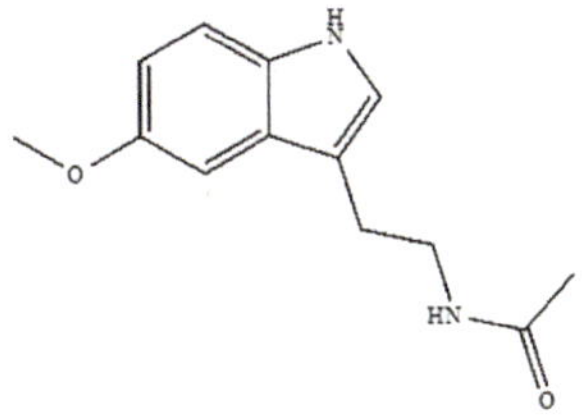
	Structure 142	10-Methoxy harmalan	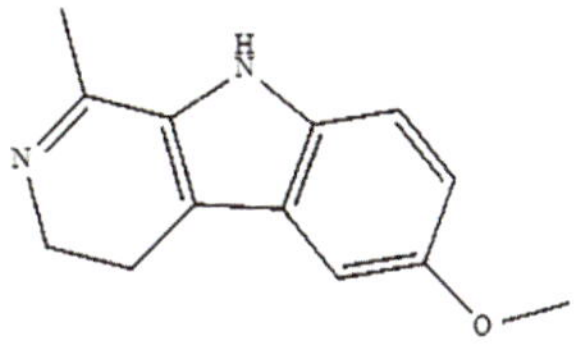

REFERENCES

Amoussa, A.M., M. Bourjot, L. Lagnika, C. Vonthron-Sénécheau, and A. Sanni. 2016. Acthaside: A new chromone derivative from *Acacia ataxacantha* and its biological activities. *BMC Complement Altern Med* 16(1): 506.

Chaubal, R., A.M. Mujumdar, V.G. Puranik, V.H. Deshpande, and N.R. Deshpande. 2003. Isolation and X-ray study of an anti-inflammatory active androstene steroid from *Acacia nilotica. Planta Med* 69(3): 287–8.

Fotso, W.G., J. Na-Iya, T.A. Mbaveng, et al. 2019. Polyacanthoside A, a new oleanane-type triterpenoid saponin with cytotoxic effects from the leaves of *Acacia polyacantha* (Fabaceae). *Nat Prod Res* 33(24): 3521–6.

Haridas, V., Z.X. Xu, D. Kitchen, A. Jiang, P. Michels, and J.U. Gutterman. 2011. The anticancer plant triterpenoid, avicin D, regulates glucocorticoid receptor signaling: Implications for cellular metabolism. *PLoS One* 6(11): e28037.

Jæger, D., C.P. Ndi, C. Crocoll, et al. 2017. Isolation and structural characterization of echinocystic acid triterpenoid saponins from the Australian medicinal and food plant *Acacia ligulata. J Nat Prod* 80(10): 2692–8.

Mahato, S.B., B.C. Pal, and A.K. Nandy. 1992. Structure elucidation of two acylated triterpenoid bisglycosides from *Acacia auriculiformis* Cunn. *Tetrahedron* 48(32): 6717–28.

Index

Note: Page numbers in *italics* indicate a figure and page numbers in **bold** indicate a table on the corresponding page.

B

P

Q

R

S

T

U

V